Management Division 2013

Core Programming Topic at the 2013 AIChE Spring Meeting & 9th Global Congress on Process Safety

San Antonio, Texas, USA
28 April – 2 May 2013

ISBN: 978-1-62748-458-9

Printed from e-media with permission by:

Curran Associates, Inc.
57 Morehouse Lane
Red Hook, NY 12571

Some format issues inherent in the e-media version may also appear in this print version.

Printed by Curran Associates, Inc. (2013)

For permission requests, please contact AIChE
at the address below.

AIChE
3 Park Avenue
New York, NY 10016-5991

Phone: (203) 702-7660
Fax: (203) 775-5177

www.aiche.org

Additional copies of this publication are available from:

Curran Associates, Inc.
57 Morehouse Lane
Red Hook, NY 12571 USA
Phone: 845-758-0400
Fax: 845-758-2634
Email: curran@proceedings.com
Web: www.proceedings.com

TABLE OF CONTENTS

(7a) Introduction to the MBA

Jon Haughton, Health & Nutrition, Corn Products International, Westchester, IL, Meagan Lewis, UOP, IL and John M. Forgac, BP Naperville Campus, BP Products North America, Naperville, IL

This is a tutorial-like presentation of the MBA-In-A-Day session. An invited speaker will give an overview of the title subject, who could be a faculty member from a regional business school or an expert practitioner.

The session will be of interest to those who want to learn about MBA programs and content: Those who may be considering enrolling in an MBA program, those who want to understand the type of information that is covered, and those who would benefit from having an overview of a typical MBA program. The speakers will direct their presentations to the interests of chemical engineers. The session will be highly of interest to young professionals: The typical MBA student has six years of experience and is 29 years old, and students with engineering or science backgrounds make up 35 to 40% of MBA programs.

(7b) Entrepreneurship and Innovation

John M. Forgac, BP Naperville Campus, BP Products North America, Naperville, IL, Meagan Lewis, UOP, IL and Jon Haughton, Health & Nutrition, Corn Products International, Westchester, IL

This is a tutorial-like presentation of the MBA-In-A-Day session. An invited speaker will give an overview of the title subject, who could be a faculty member from a regional business school or an expert practitioner.

The session will be of interest to those who want to learn about MBA programs and content: Those who may be considering enrolling in an MBA program, those who want to understand the type of information that is covered, and those who would benefit from having an overview of a typical MBA program. The speakers will direct their presentations to the interests of chemical engineers. The session will be highly of interest to young professionals: The typical MBA student has six years of experience and is 29 years old, and students with engineering or science backgrounds make up 35 to 40% of MBA programs.

(7c) Accounting, Finance, and Statistics I

Jon Haughton, Health & Nutrition, Corn Products International, Westchester, IL, Meagan Lewis, UOP, IL and John M. Forgac, BP Naperville Campus, BP Products North America, Naperville, IL

This is a tutorial-like presentation of the MBA-In-A-Day session. An invited speaker will give an overview of the title subject, who could be a faculty member from a regional business school or an expert practitioner.

The session will be of interest to those who want to learn about MBA programs and content: Those who may be considering enrolling in an MBA program, those who want to understand the type of information that is covered, and those who would benefit from having an overview of a typical MBA program. The speakers will direct their presentations to the interests of chemical engineers. The session will be highly of interest to young professionals: The typical MBA student has six years of experience and is 29 years old, and students with engineering or science backgrounds make up 35 to 40% of MBA programs.

(7d) Accounting, Finance, and Statistics II

John M. Forgac, BP Naperville Campus, BP Products North America, Naperville, IL, Meagan Lewis, UOP, IL and Jon Haughton, Health & Nutrition, Corn Products International, Westchester, IL

This is a tutorial-like presentation of the MBA-In-A-Day session. An invited speaker will give an overview of the title subject, who could be a faculty member from a regional business school or an expert practitioner.

The session will be of interest to those who want to learn about MBA programs and content: Those who may be considering enrolling in an MBA program, those who want to understand the type of information that is covered, and those who would benefit from having an overview of a typical MBA program. The speakers will direct their presentations to the interests of chemical engineers. The session will be highly of interest to young professionals: The typical MBA student has six years of experience and is 29 years old, and students with engineering or science backgrounds make up 35 to 40% of MBA programs.

(31a) Marketing

Jon Haughton, Health & Nutrition, Corn Products International, Westchester, IL and Meagan Lewis, UOP, IL

This is a tutorial-like presentation of the MBA-In-A-Day session. An invited speaker will give an overview of the title subject, who could be a faculty member from a regional business school or an expert practitioner.

The session will be of interest to those who want to learn about MBA programs and content: Those who may be considering enrolling in an MBA program, those who want to understand the type of information that is covered, and those who would benefit from having an overview of a typical MBA program. The speakers will direct their presentations to the interests of chemical engineers. The session will be highly of interest to young professionals: The typical MBA student has six years of experience and is 29 years old, and students with engineering or science backgrounds make up 35 to 40% of MBA programs.

(31b) Strategy and Economics

John M. Forgac, BP Naperville Campus, BP Products North America, Naperville, IL, Meagan Lewis, UOP, IL and Jon Haughton, Health & Nutrition, Corn Products International, Westchester, IL

This is a tutorial-like presentation of the MBA-In-A-Day session. An invited speaker will give an overview of the title subject, who could be a faculty member from a regional business school or an expert practitioner.

The session will be of interest to those who want to learn about MBA programs and content: Those who may be considering enrolling in an MBA program, those who want to understand the type of information that is covered, and those who would benefit from having an overview of a typical MBA program. The speakers will direct their presentations to the interests of chemical engineers. The session will be highly of interest to young professionals: The typical MBA student has six years of experience and is 29 years old, and students with engineering or science backgrounds make up 35 to 40% of MBA programs.

(31c) Operations, Supply Chain, and Risk Management

Jon Haughton, Health & Nutrition, Corn Products International, Westchester, IL, Meagan Lewis, UOP, IL and John M. Forgac, BP Naperville Campus, BP Products North America, Naperville, IL

This is a tutorial-like presentation of the MBA-In-A-Day session. An invited speaker will give an overview of the title subject, who could be a faculty member from a regional business school or an expert practitioner.

The session will be of interest to those who want to learn about MBA programs and content: Those who may be considering enrolling in an MBA program, those who want to understand the type of information that is covered, and those who would benefit from having an overview of a typical MBA program. The speakers will direct their presentations to the interests of chemical engineers. The session will be highly of interest to young professionals: The typical MBA student has six years of experience and is 29 years old, and students with engineering or science backgrounds make up 35 to 40% of MBA programs.

(31d) Leadership, People Management, and Organizations

John M. Forgac, BP Naperville Campus, BP Products North America, Naperville, IL, Meagan Lewis, UOP, IL and Jon Haughton, Health & Nutrition, Corn Products International, Westchester, IL

This is a tutorial-like presentation of the MBA-In-A-Day session. An invited speaker will give an overview of the title subject, who could be a faculty member from a regional business school or an expert practitioner.

The session will be of interest to those who want to learn about MBA programs and content: Those who may be considering enrolling in an MBA program, those who want to understand the type of information that is covered, and those who would benefit from having an overview of a typical MBA program. The speakers will direct their presentations to the interests of chemical engineers. The session will be highly of interest to young professionals: The typical MBA student has six years of experience and is 29 years old, and students with engineering or science backgrounds make up 35 to 40% of MBA programs.

Overcoming Regulatory Challenges In Chemical Management - Inherent Opportunities for the Chemical Industry

Jacqueline Sibblies, PE
Environmental Compliance Consulting
P.O. Box 5013, Somerset, NJ 08873
sibbliesj@gmail.com
(908) 531-4029

The first decade of this century has undoubtedly ushered in multiple changes and new beginnings in the way we live and do business locally, nationally and internationally. Several newly implemented and updated regulations and policies have resulted in profound changes in the operations of various industrial sectors. Among the industries significantly affected by new regulations and policies is the chemical manufacturing industry, uniquely subject to multiple regulatory and policy changes which, while not implemented with much public fanfare, are nevertheless sparking creative, new industry paradigms.

The industry is required to comply with extensive environmental and health/safety (EHS) requirements at local, national and global levels. Many of these regulations are administered by different regulatory bodies, and many have overlapping requirements. For example, in the United States (US), chemical companies are subject to on-site chemical handling/safety rules by two agencies - the Environmental Protection Agency (EPA) through several regulations including the Clean Air Act requiring the Risk Management Plan (RMP), and the Occupational Safety and Health Administration (OSHA) through requirements like Process Safety Management (PSM). In addition to these federal (or national) environmental, health and safety (EHS) requirements, the industry is also subject to global policy requirements such as the Strategic Approach to International Chemicals Management (SAICM) administered by the United Nations Environmental Programme (UNEP). All of these requirements include components on safe chemical handling to preserve worker safety and prevent or mitigate releases to the environment. In essence, chemical companies are obligated to comply concurrently with numerous chemical management standards by multiple entities. This is considered burdensome by most and a detriment to the business survival by many industry stakeholders.

Yet despite this perceived threat to business profitability, fulfilling these regulatory and policy requirements offers inherent opportunities for business growth and development. Indeed, some industry leaders like Dow Chemical Company (DOW) and BASF have already seized the opportunity to develop sustainability tools and models while devising a comprehensive system for fulfilling the varied regulatory requirements. In addition to addressing social, ecological and business profitability concerns, sustainability explorations could lead to discovery of alternative raw materials and applications, new product development and improvement in process operations, all positively affecting business performance. Results could extend into the global arena through the forging of strategic alliances to overcome national barriers that may inhibit ease of international compliance. Furthermore, these alliances could lead to improved global presence and established international branding. This paper will summarize key regulatory challenges in Environmental Health and Safety (EHS)/chemical management faced by the chemical industry and explore opportunities for business advancement through overcoming these challenges, citing industry cases.

Myriad Regulatory and Policy Requirements

For the past four decades, the chemical industry has been subjected to growing regulatory obligations. These have been promulgated primarily in the form of new regulations and significant amendment to existing regulations. The industry, for its part, has sought to lend its voice to the rule making process in an effort to mitigate any adverse effects to its operations that final decisions might cause. These efforts, however, have been tempered from time-to-time by various mishaps and catastrophic events which precipitate mounting public concerns, sometimes outrage.

From the 1976 Seveso disaster in Italy (which was catastrophic to livestock and other animal life), to the 1984 Bhopal India accidental toxic release (which caused thousands of deaths), to the 2005 Texas City, Texas refinery explosion and fire (which claimed fifteen lives), chemical accidents have served to frame a perception that the chemical industry is inherently dangerous. These accidents not only brought the responsible facility into high visibility and under subsequent intense scrutiny, but led to repeated critique of the entire industry. They resulted in the enactment of regulations and policies mandating preventive safety measures not only from local or national jurisdiction, but also from regional environmental regulatory authorities and international stakeholder consortia.

The 1976 incidents prompted the European Commission (EC) to issue the Seveso Directive, legislation governing prevention and mitigation of chemical accidents. The Bhopal incident generated debate in the United States Congress, resulting in the Emergency Planning and Community Right-to-Know (EPCRA) legislation, focused on emergency response and public education. There were further environmental, health and safety evaluations and discussions which led to the 1990 amendment of the Clean Air Act which required companies to develop a risk management plan. Over subsequent years, other regulatory and policy requirements effected for the purpose of having the industry practice better risk management in chemical handling. The EC and the US governmental agencies were not the only bodies which proceeded with these types of regulatory action; the United Nations (UN) also developed internationals standards. Table 1 (Appendix A) gives a summary of current regulations that contain chemical management requirements. Figure 1 illustrates the network correlation.

Fig.1. – Interconnection of National, Regional and Global Chemical Management Rules

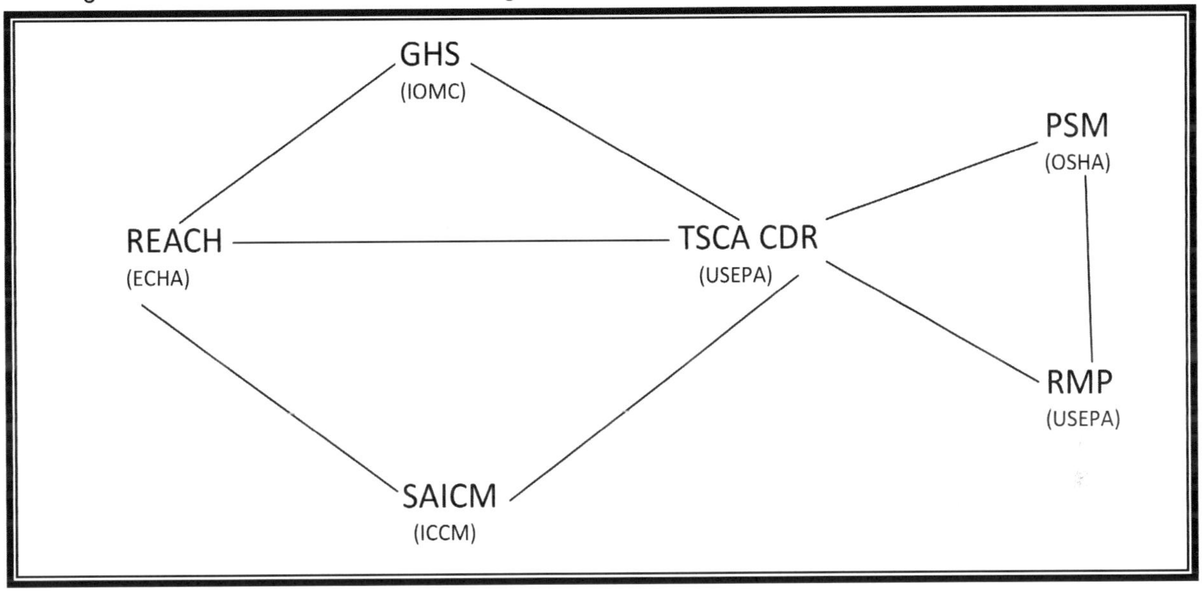

Challenges to Compliance

The Chemical industry had growing concerns about the feasibility of complying with all applicable regulations within reasonable economic limits. The financial cost of compliance was thought to be overwhelming.

For large and some midsized companies with potential capabilities for meeting compliance with in-house resources, it was felt that the cost of equipment and technological tools necessary to meet compliance would be disproportionately high. They also cited the detraction from production that additional labor-hours would require to fulfill requirements like testing, monitoring, record keeping and reporting – adversely affecting business profitability. Smaller companies, perceiving very little or no in-house compliance resource capabilities would have to rely on consulting assistance which meant more cost to a usually already lean budget.

The complexities of some of the regulations, it is safe to say, were even more disconcerting to many stakeholders in the industry value chain. As figure 1 above shows, the same agency has enacted chemical management requirements in separate regulations. Further, in the United States multiple agencies (e.g. the EPA and OSHA) have administered regulations with similar requirement for the industry. On top of that, there might also be unique state, county and city regulations that some companies have had to comply with. Small and mid-sized companies have needed assistance to decipher pertinent regulatory requirements from a multi-layered regulatory system.

To compound matters, each set of regulations established a reporting schedule, which was seldom in sync with any other set of regulations. So there was the potential for having several compliance schedules to comply with in any given calendar year. Missed deadlines could lead to a non-compliance citation, which came with a financial penalty. Companies needed a management system to manage chemical management compliance.

In addition to the direct costs associated with fulfilling regulatory requirements, the chemical industry had greater concern about the far reaching effects that these cost had on the business of doing business. A major threat to business performance in one geographic location is always competition from another geographic location. US chemical companies voiced strong opposition to many EPA regulatory requirements on the grounds that the addition cost of compliance measures added to the cost of production and that would force a product price increases, thus undermining the ability to adequately compete with companies from other geographic locations not subject to similar requirements. Alternatively, if product prices were kept at competitive levels, a company's profit margin would be undermined, posing a threat to its economic standing or even its very survival.

Compliance Strategies – Inherent Opportunities

However, demand for products from the chemical industry has been steadily rising, even through the great recession of 2008. New technological innovations brought new uses for chemicals and improved efficiency in products, creating new business opportunities for the industry. The public may ask for stringent regulations that protect its health and safety and also preserve the environment, but it also wants the highest quality and most technologically advanced products that the industry can deliver. So the question that key industry figures started grappling with was, "How can we meet stakeholder demands while simultaneously meeting the company's business goals and more importantly, shareholder expectations?"

Concurrently, as effective dates for compliance became imminent, many companies, particularly industry leaders, started taking a critical look at compliance requirements, and formulating strategies by which to achieve them. The exercise uncovered a key fact - although regulations are many and varied, requirements were found to not be mutually exclusive. As Table 1 shows, there is appreciable overlap of chemical management (and EHS) requirements not only with intra-agency rules (like the EPA's), but inter-agency rules and inter-regional and international standards as well. This scenario presented the opportunity for a company to devise a comprehensive compliance program. The chemical industry's perception of a burdensome regulatory system slowly started to change. A paradigm shift started developing around the turn of this century.

The new thinking gave rise to the development and application of environmental management systems (EMS) tools and models developed by EMS teams. Most large companies comprised their EMS teams from internal company labor pools, with team members taken from departments and divisions directly affecting chemical management and with personnel having the knowledge and experience with affected processes.

It also ignited wider interest in industry standards developed by industry and stakeholder consortium, for example, the chemical industry's own Responsible Care Initiative and the International Organization for Standardization (ISO) guidelines for environmental management (ISO 14001). With these available mechanisms, companies did not have to reinvent the wheel when exploring strategies for chemical management compliance; they already had these established platforms on which to build.

Many large, international companies had already been subscribing to ISO and Responsible Care principles from the 1990's. Both initiatives are voluntary; members of the industry's supply chain globally had been encouraged to adopt these programs and incorporate the principles into their standard operations, to bring about improvements in health, safety and environmental performance and to foster public trust through transparency in communication.

In adapting these kinds of initiatives to accommodate compliance with varied chemical management regulatory requirements, companies had the opportunity to undertake critical review of their entire operation (processes and equipment) and standard operating procedures (SOP's), including all activities in the supply chain. This enabled the opportunities to identify possible areas of redundancies or inadequacies and devise appropriate means of improvement. It also allowed for the possibilities of new discoveries, whether of equipment or device application, processes, intermediate product application or areas for closer inter-department collaboration.

In essence, this exercise forces companies to evaluate what works and what does not; what may benefit from improvement and what may need to be illuminated, from technical operations to business operations. And in the process, they can develop a comprehensive program for fulfilling chemical management regulatory requirements – testing and risk assessment of chemicals, recording keeping of data, reporting required information, filing certification or permit applications – i.e. applying a holistic, sustainable approach.

This approach sustainable approach includes principles of sustainability, which seek to optimize economic benefits to an organization's business, while protecting the environment or eco-system and enhancing human wellbeing. And these are also the ultimate objectives of chemical management regulations, policies and initiatives, to manage chemicals in the supply chain in such a manner that none of these areas are compromised.

BASF, a global industry giant, is one stellar example of a company that employed a holistic approach to chemical management, business operation, environmental, health and safety compliance and ecological preservation. The company developed an analytical model called the SEEBALANCE (illustration below in Fig. 2) which can be used to analyze any of the pillars of sustainability, impacts and cost, and also correlation between any combinations of the three pillars.

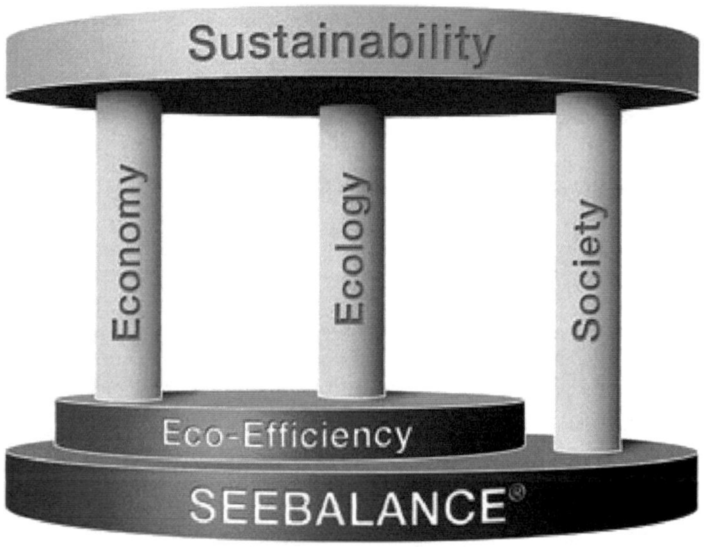

Fig. 2 – Illustration of BASF's Innovative Comprehensive Analytical Tool

Overwhelming public opinion is that regulations are necessary to help attain this tripod balance. And companies are very much aware of the value of meeting public expectation – the result could be the establishment of a lucrative brand or the demise of an unsuccessful one – directly affecting economic performance and the bottom line.

Various institutions and regulatory bodies offer award incentives to companies who attain high levels of achievement in areas such as environmental preservation, health and safety record, product development that enhances ecological preservation (green chemistry) and operational changes that minimizes a company's or chemical's carbon footprint. These incentives also encourage companies to evaluate their overall operation for areas of and opportunities for national and global recognition.

Following are some notable industry achievements and recognition:

- Dow Chemical was honored and a 2012 top 100 innovator by Thompson Reuters, noting "The company connects chemistry and innovation with the principles of sustainability ..."*
- Henkel, Serbia and SAFECHEM Europe were two of the 2012 recipients of the Global Chemical Leasing Award – chemical leasing is a service-oriented business model that shifts the focus from increasing sales volume of chemicals towards a value added approach

Chemical companies may also seek to forge synergistic collaborations and strategic alliances as part of their strategy to accomplish chemical management objectives at optimal benefit to business operation. It is from on such collaboration between BASF and various academic research institutes that the SEEBALANCE analytical was developed. Strategic alliances could make major contributions in a product leasing initiative or, more directly, the development of a corporate chemical management compliance system that facilities worldwide can use as a central resource of information on compliance.

Conclusion

The safe management of chemicals has for decades been regarded as an essential policy of operation by all stakeholders in the chemical industry: manufacturers and distributors, intermediate and end product users, the public and regulators. But differences of opinion as to the most viable and balanced way of achieving an outcome acceptable to all stakeholders had persisted until the turn of the 21st century, when a paradigm shift started spreading across the industry. Companies started recognizing the opportunities in developing a comprehensive chemical management program which would address compliance requirements of multiple regulations, while supporting the bottom-line.

A recent survey conducted by *Chemical Watch* indicates that this trend is expected to continue over the next five years, with a growing number of mid-sized and small companies adopting the comprehensive approach. The future promises to be quite sustainable.

Reference Sources

Chemical Watch, 24 January, 2013
www.ICC-Chem.org
*www.dow.com
www.unep.org
www.basf.com

Appendix A

Table 1. Regulations and Policies Containing Chemical Management Components

Administrating Body	Regulations/Policies	Chemical Management Requirements and Standards
United States Environmental Protection Agency (USEPA)	Clean Air Act (CAA)	Risk Management Plan (RMP) – Planned changes to improve safety in chemical handling
	Toxic Substances Control Act (TSCA)	Testing and assessment of chemicals and reporting of data on risks and hazards from use and exposure to them; Certification of information
	TSCA Chemical Data Reporting (CDR)	Periodic update of information on chemicals reported under TSCA;
Occupational Safety and Health (OSHA)	Process Safety Management (PSM) of highly hazardous chemicals	Comprehensive management program for preventing / mitigating hazards associated with processes using highly hazardous chemicals;
European Chemical Agency (ECHA)	Registration, Evaluation, Authorization of Chemical Substances (REACH)	Testing of chemicals and assessment of data to determine risk; Develop and recommend appropriate risk management measures; Register
Inter-organizational Program for the Sound Management of Chemicals (IOMC)	Global Harmonized System (GHS) of chemical classification	Internationally accepted system for the classification and labeling of chemicals
International Conference on Chemical Management (ICCM)	Strategic Approach to International Chemicals Management (SAICM)	Framework to develop and implement sound chemical management practices globally

Risk-Based Approach to Building Designs and Decision Making

John N. Dyer P.E., Anay P. Raibagkar P.E., Anthony G. Sarrack (Baker Engineering and Risk Consultants, Inc.)

Introduction

Accidental releases of flammable and toxic materials from hazard sources of different sizes ranging from small leaks to full-bore pipe ruptures can represent a wide range of impacts to buildings. Facility siting for petrochemical and chemical plants is required to cover credible accident scenarios. Results of such studies often include catastrophic consequences that may be impractical to eliminate through structural designs. Rather than designing buildings for a high level of protection against the most severe credible events, a risk-based approach provides a practical and a cost effective approach to maximize safety.

Two fundamental principles are required to be met for a design to qualify as "Risk-Based". One principle is that each occupant of the building has a minimum level of protection. This principle is achieved by designing the building in such a way that each occupant's individual risk is within defined tolerance criteria. The other principle is that the building is designed in a way such that it reduces the overall societal risk to a level that is as low as practical. This principle is achieved by comparing costs of enhanced designs against the safety benefit they afford and implementing as much mitigation as is practical. Employing these fundamental principles ensures that cost is optimized for a given risk level, allowing risk mitigation resources to be optimized.

This paper outlines a methodology to perform a risk-based building design, accounting for predicted impacts of a wide range of credible accidents, associated frequencies, and building occupancy. The necessary input data for this type of design is calculated in a Quantitative Risk Assessment. The proposed methodology will help designers establish practical structural requirements to minimize risk while achieving cost effective designs.

Building Blast Damage and Occupant Vulnerability

The expanding pressure wave resulting from an accidental explosion will load objects in its path with a relatively uniform pressure applied to all exposed surfaces, as illustrated in Figure 1. The sides of a structure facing the source of the explosion will cause the wave to reflect, increasing the magnitude of the overpressure on those surfaces.

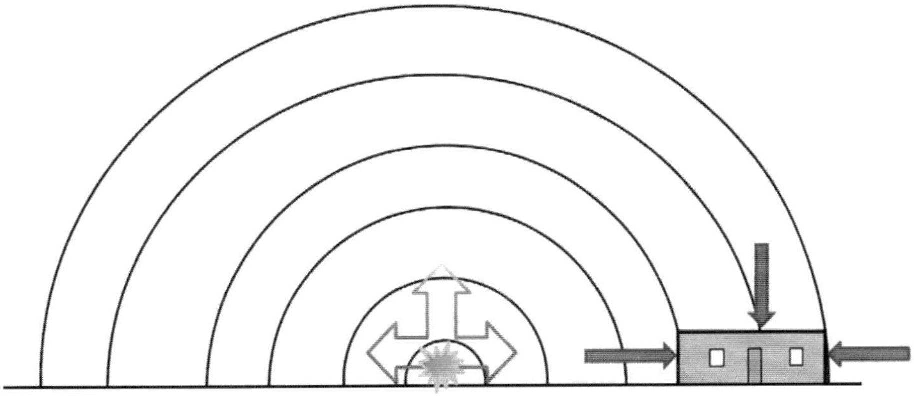

Figure 1. Pressure Wave Loading of Building

The explosion pressure may be considered to be applied instantaneously and to decay to the ambient pressure linearly, as illustrated in Figure 2. The time that the overpressure is applied on a building's surfaces is usually a small fraction of a second. The impulse of the blast load is defined as the integral of the load function. For the idealized triangular load history illustrated in Figure 2, the impulse may be approximated as the area of the triangle. The intensity of a blast load on a building surface may be expressed in terms of its peak pressure and duration or by its peak pressure and impulse.

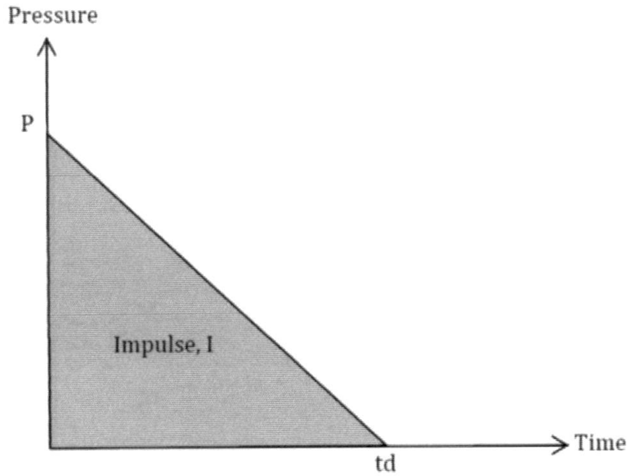

Figure 2. Idealized Overpressure History

The blast will cause structural elements on the loaded surfaces of a building to respond in an oscillatory fashion. Components that are not catastrophically damaged, will reach a peak deflection followed by a rebound phase. The oscillations will become smaller with time until they stop. The level of protection that a building affords its occupants is typically dependent on its ability to resist blast loads without sustaining significant damage. Therefore, estimating the building damage level (BDL) caused by explosions is important.

Occupant vulnerability (OV) is defined as the fraction of the building population who will sustain fatal injuries. Since the OV is connected to the damage sustained by the building, establishing a BDL categorization is useful to assess building damage and occupant vulnerability. Table 1 provides a summary of typical building damage levels with a brief description of structural component damage.

Table 1. Building Damage Level Categories [1]

BDL	Damage Description
1	Minor damage to roof and walls
2	Moderate or lower damage to roof and load-bearing components
2.5	Localized Major damage to roof or walls. Moderate damage to other components
3	Major damage to load-bearing elements or roof
4	Failure of load-bearing elements or roof collapse

The two lesser damage levels (BDL 1 and BDL 2) are typically associated with minor to moderate component damage particularly to the sides of the building facing the explosion, as illustrated in Figure 3. BDL 1 is used in instances where very low damage is sustained, and BDL 2 corresponds to cases in which some building components may have sustained moderate damage. When a building sustains BDL 1 or BDL 2, personnel inside the building are not considered to suffer life-threatening injuries.

Figure 3. Minor Building Damage

BDL 2.5 is used to represent instances where major building damage is sustained including some structural failures on the walls. BDL 2.5 is associated with low (not negligible) occupant vulnerability, since life-threatening injuries could typically only result to personnel located near the most heavily damage areas, as those shown in Figure 4.

BDL 3 and BDL 4 are associated with widespread structural damage that could potentially compromise the integrity of the entire building. Occupant vulnerability is typically significant [1] at BDL 3 and 4. In particular, BDL 4 is used to represent collapse of the roof, which can result in a large portion of the building population sustaining fatal injuries (0.5 < OV < 1.0).

Figure 4. Major Building Damage

Building Blast Assessment

Designing buildings to resist explosion blast loads requires structural methods that allow quantifying the level of building damage and estimating occupant vulnerability. Since the severity of a blast load cannot be fully assessed without knowing its peak pressure and impulse, the structural method should take into account not only the pressure but also its duration. Dynamic structural analyses will allow taking into account the duration of the blast load. In order to estimate structural damage, the elastic-plastic behavior of the structure must be modeled. Since more than one explosion case is often necessary to be evaluated, the structural methodology should preferably be based on simplified analyses to avoid expensive structural calculations. The Single-Degree-of-Freedom (SDOF) methodology allows engineers to perform relatively inexpensive calculations with a reasonable level of precision. The SDOF methodology can also be used to construct component damage curves based on pressure-impulse (P-i) combinations that result in equal component response. The component damage curves can be combined to construct building damage Pressure-impulse diagrams as the one shown in Figure 5. The SDOF methodology is a widely accepted analysis technique and is described in several references [2, 3].

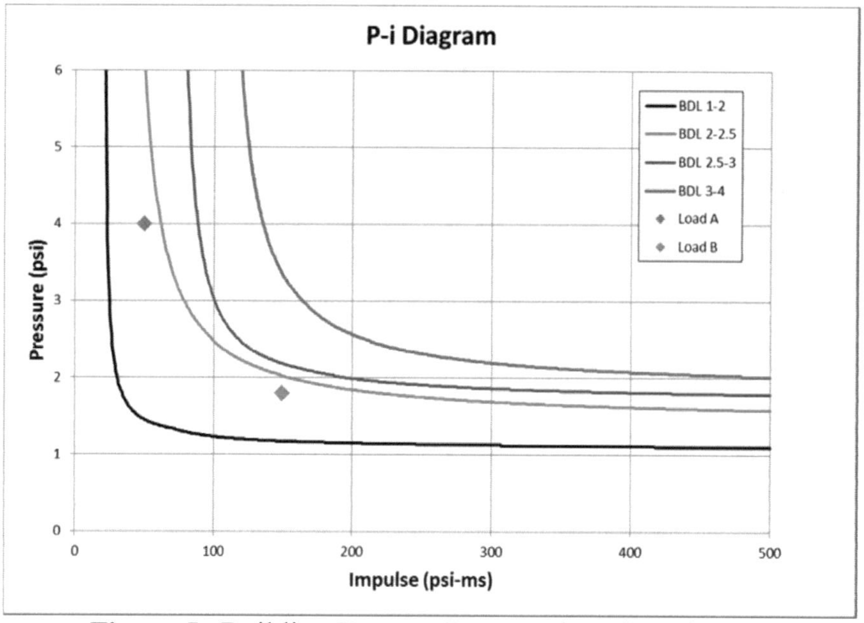

Figure 5. Building Damage Pressure-impulse Diagram

The P-i curves represent boundaries of BDL regions. Blast loads can be plotted on the P-i diagram to perform graphical blast assessments. Since the curves represent blast loads that result in a specific damage, a blast load represented by a P-i point will result in higher or lower damage than that represented by the curve depending on their relative position. For example, the two blast loads with different pressure and impulse shown in Figure 5 would cause the same damage level since they are bound by the same BDL curves. P-i diagrams are particularly useful when damage from multiple blast loads need to be assessed, since evaluating multiple component under numerous blast loads can be costly.

Occupant Vulnerability

Estimating occupant vulnerability (OV) is the quantification of loss of human life inside a building from an explosion. OV multiplied by the number of people in a building is typically the "consequence" used in calculating explosion risk. The main driver of OV is BDL [1] and can be estimated by calculating damage on the roof and walls, as illustrated in Figure 6.

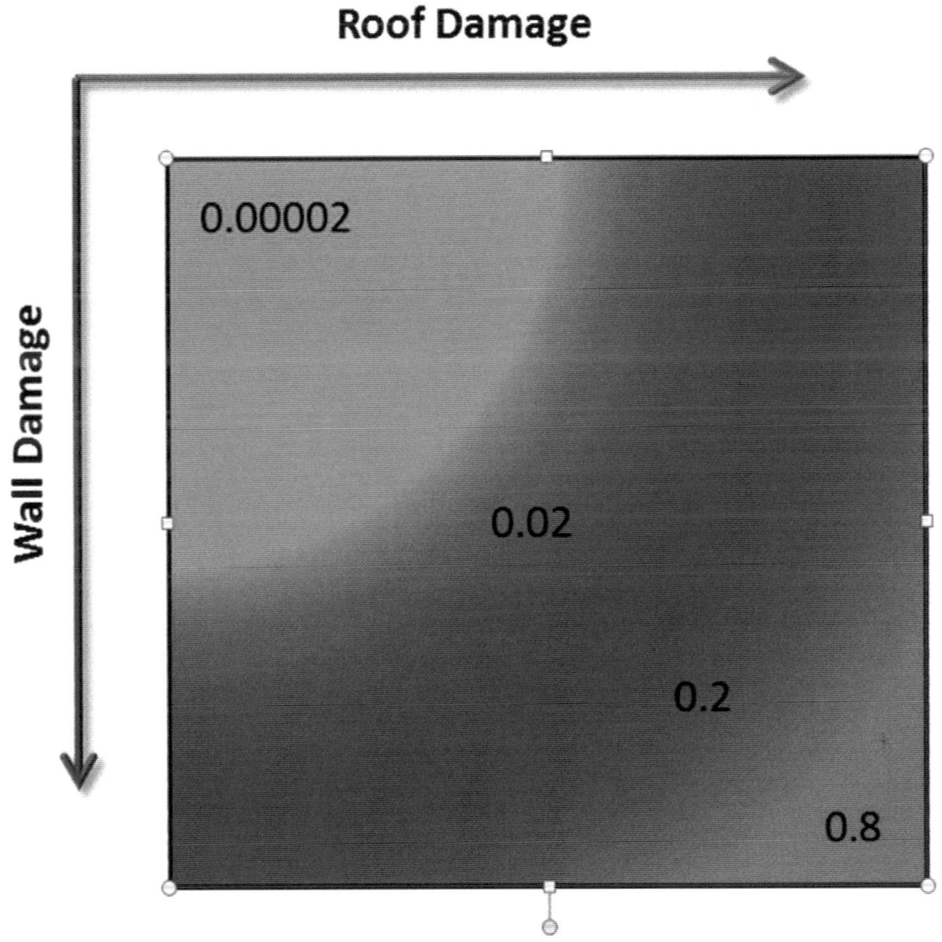

Figure 6. OV as a Function of Structural Damage (typical)[1]

However, factors such as the distribution of the population inside the building, the amount of window coverage on exterior walls, the structural system of the building, and the weight of the construction can influence the OV [1].

[1] This is a generic figure. The actual values of OV will depend on several factors such as type of construction, window coverage on walls, perimeter population in the building etc. A detailed methodology for calculating OV is given in [1].

Table 2. Typical Occupant Vulnerability Values (for Each BDL)

BDL	Typical Occupant Vulnerability	
	Steel Frame with Metal siding	**Concrete Frame with masonry walls**
1	0.00002	0.00001
2	0.00008	0.00005
2.5	0.015	0.02
3	0.2	0.3
4	0.5	0.95

The consequences in terms of fatalities can be calculated for a building for each explosion scenario by assessing building damage and the corresponding occupant vulnerability which is multiplied by the occupancy of the building to estimate the total number of fatalities. Therefore, the consequence of an explosion (for a particular scenario) can be calculated as

$$\text{Consequence}_{\text{bldg-exp}} = \text{OV}_{\text{bldg-exp}} \text{ x } \text{Pop}_{\text{bldg.}}$$

Where,

$\text{OV}_{\text{bldg-exp}}$ – Occupant Vulnerability corresponding to the BDL

$\text{Pop}_{\text{bldg.}}$ – Average occupancy in the building

Risk Assessment[2]

Risk is defined as the product of consequence (fatalities / event) and frequency (events / year) and is presented in terms of fatalities per year. Therefore, risk (for a particular scenario) can be calculated as

$$\text{Risk}_{\text{scenario}} = \text{F}_{\text{scenario}} \text{ x } \text{Consequence}_{\text{scenario}}$$

The frequency of each explosion is calculated by multiplying the release frequency by the applicable conditional probabilities of factors affecting the likelihood of the explosion such as

- Equipment failure rates
- Weather conditions
- Wind direction
- Ignition probability, and
- Time slot probability

Therefore, the frequency of the explosion scenario can be calculated as

[2] The risk assessment method given in this paper is for explosion risk only. The method can be extended to other hazards such as toxic and fires.

$$F_{scenario} = R_{eqpt} \times C_{eqpt} \times P_{weather} \times P_{wind} \times P_{ignition} \times P_{time}$$

When blast loads are relatively low at the location of interest, designing a building to sustain low or moderate damage is typically enough to ensure that the OV is negligible, and risk will be typically low. However, when a building is potentially subjected to high overpressures, as illustrated in Figure 7, the structural requirements to mitigate higher than moderate (BDL 2) damage can be excessive making the building design cost prohibitive.

In order to achieve a cost-effective design and provide an adequate level of protection under relatively high blast loads, the design of the building must take the frequency of the explosion scenarios and be based on achieving a tolerable risk level. Thus, the total societal explosion risk for a building can be calculated by adding risk from all postulated scenarios as shown below.

$$\text{Societal Explosion Risk} = \sum \text{Risk}_{scenario} = \sum (F_{scenario} \times \text{Consequence}_{scenario})$$

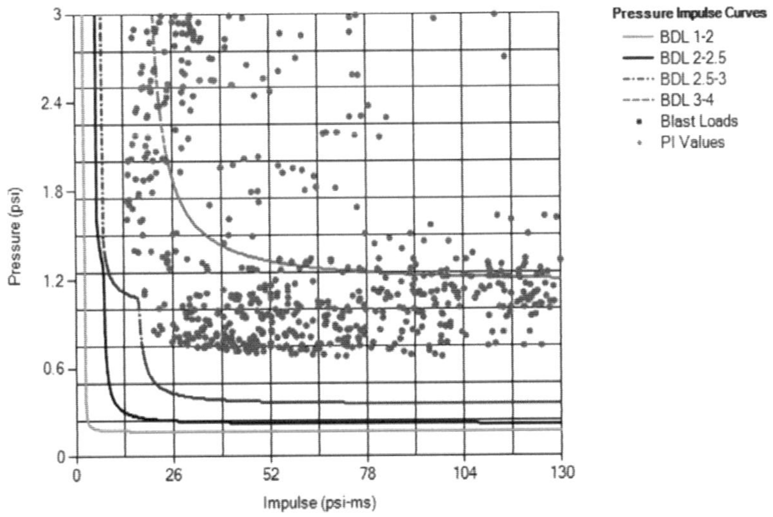

Figure 7. P-i Diagram with All Credible Events

Risk-Based Structural Design

The explosion risk methodology summarized in this paper can be effectively used in selecting a cost effective design as indicated in Figure 8. The figures show explosion risk vs. additional building costs for four design options. The 'Original' design can be considered as a baseline (cheapest design that will satisfy the local building codes only) design with no extra resources spent for the purposes of explosion risk mitigation. The remaining three options, give considerably lower risk compared to the 'Original' baseline design. However, the chart on the right indicates that additional resources will be required to lower the risk below the tolerable level. As indicated in the 'Explosion Risk' chart, the risk associated with the 'BRM' option is just below the tolerable limit of 8.0E-04 and the corresponding cost of $360,000 is the lowest among the three options under the tolerable risk. Thus, in this particular example, the BRM option can be considered as the most effective risk mitigation solution and can be selected for construction.

However, this choice is based on explosion risk only. Similarly, risk from toxic releases and fire hazards may be present and the BRM option may not be a suitable choice. Thus, a similar risk based method should be applied in selecting a cost effective design option and making decision for effectively allocating resources and maximize safety.

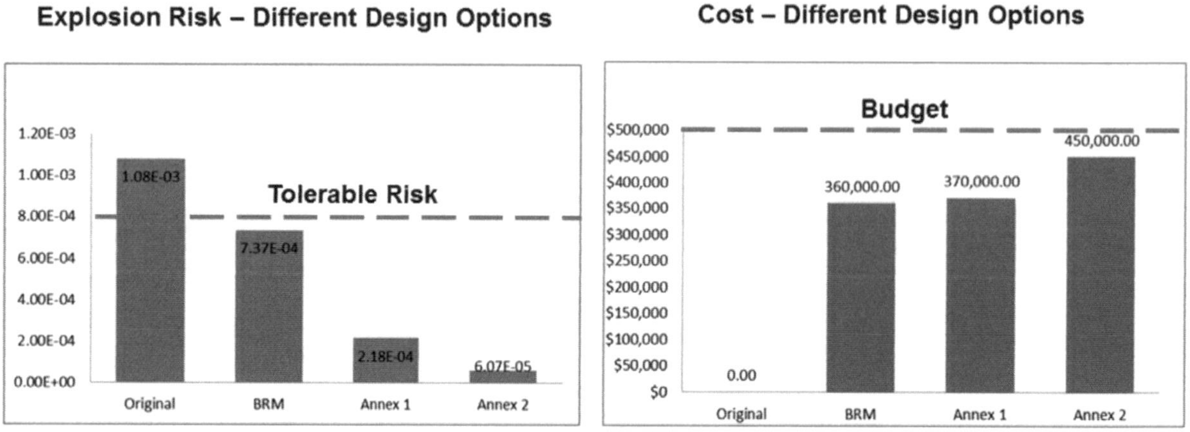

Figure 8 Comparison between cost of a building vs. Risk to building occupants

References

1. Oswald, Charles J. and Quentin A. Baker (1999), "Vulnerability Model for Occupants of Blast Damaged Buildings," *Prepared for Presentation at the 34th Annual Loss Prevention Symposium, March 6-8, 2000, Session 3.*
2. Biggs, J.D (1964), "Introduction to Structural Dynamics," McGraw-Hill Publishing Company, New York.
3. ASCE, ASCE Task Committee on Blast Resistant Design, Design of Blast Resistant Buildings in Petrochemical Facilities, American Society of Civil Engineers, 2nd Edition, New York, 2010.

(32c) EPA's Greenhouse Gas Reporting Program: Lessons Learned for Onshore Petroleum and Natural Gas Production Operations

Karin Ritter, American Petroleum Institute, Washington DC, DC, Miriam Lev-On, The LEVON Group, Thousand Oaks, CA and **Theresa Shires**, URS Corporation, Austin, TX

The Environmental Protection Agency (EPA) mandatory greenhouse gas (GHG) reporting program applies to facilities and operations whose annual GHG emissions exceed 25,000 tonnes of carbon dioxide equivalent. For the Petroleum and Natural Gas Systems sector, 40 CFR Part 98 Subpart W lists the requirements for quantification and reporting of GHG emissions. The methodologies specified for the eight industry segments within the sector, range from application of specified emission factors, engineering calculations, and direct emissions measurements. Notably, the Onshore Petroleum and Natural Gas Production segment is comprised of a large number of emission sources within "facilities" that cover multiple operations within a single hydrocarbon basin. These requirements necessitate installation of proper instrumentation, robust data collection practices, and Information Technology (IT) systems that enable a smooth interface with EPA's electronic greenhouse gas reporting tool (eGGRT).

This paper will address the lessons learned from the initial phases of reporting for the Petroleum and Natural Gas systems and will emphasize the unique aspects of reporting GHG emissions for onshore oil and natural gas production operations. The discussion will address the need to balance reliable information and reporting burden and will highlight industry's commitment to continuous improvement. It will address how reporting provides companies a means to highlight altering practices like venting and flaring minimization, and reduction of environmental impacts of operations. Such systematic data collection, as mandated by the EPA GHG reporting program, may not only help target areas for future reductions but also demonstrate significant voluntary progress toward continually 'greener' operations.

(32d) Trials and Tribulations in Carbon Foot Printing

Dr. Sunil Hangal, Environmental Compliance Assurance Management (ECAM), Bedminster, NJ

Carbon foot printing is a determination of the total greenhouse gas emissions from an organization, product or event. This presentation describes the trials and tribulations in obtaining accurate and verifiable total greenhouse gas emissions through the complete lifecycle of a product. Complete lifecycle emissions include all direct and indirect greenhouse gas emissions including carbon dioxide, nitrous oxide, and hydrofluorcarbons (HFCs) through the raw material to disposal after end use of a product.

In this study, greenhouse gas emissions were estimated for a textile fabric sold wholesale to customers. This company identified two popular products as a test case to obtain carbon footprint data. These two products were high end product with higher recycled fiber content and a product line with 100 % natural fiber origination in the US. The product lifecycle boundaries were agricultural emissions, manufacturing, transportation and storage and use of product. The goal was to obtain carbon dioxide equivalent emissions per unit measure of cloth that could be provided to end use customers.

Detail lifecycle planning was done and emission was estimated using known emission factors and activity data across the lifecycle. There were several issues that affected the accuracy of data such as variability of fertilizer use in growing cotton, reliability of some emission factors, and apportionment of activity for transportation sources, amongst others. This presentation will also discuss practical issues in working with suppliers and vendors down the supply chain in order to obtain accurate and verifiable activity data. In addition data rollout planning and difficulties with benchmarking will be discussed.

Title: A Corporate Career Supports Entrepreneurship
Speaker: Deborah L. Grubbe, PE, CEng.
Date: Monday, 29 April 2013

Abstract: It is possible to enjoy a long career in the corporate world and to achieve success as an engineering entrepreneur. In fact, if managed well, a corporate career serves an essential role in assisting the future business owner in these areas: 1) gaining essential engineering and business experience, 2) establishing credibility in your future field of practice, 3) building your global business network and 4) obtaining the needed credentials and certifications to ensure a fast start.

In 2009, at the height of the recent economic downturn, Deborah Grubbe, after a 30+ year distinguished career with multinationals, decided to fulfill a long time dream and start her own global business. Her business results have wildly exceeded expectations, and she now has a global clientele. Let's quickly review how she was able to fulfill her dreams.

First, a 30 year corporate career will allow you to gain essential engineering experience by virtue of time, temperature and applied energy. If you are alert, you will also learn about how business is done, the role of the various functions in a successful business, and some basics about human resources, business law and finance. If you are able to have an assignment in these functions, you will learn even more. When you go to start your own business, this knowledge becomes essential, because the only person you have to rely in is yourself!

A long corporate career also allows you to build on your accomplishments over time, and thus, over time, build your credibility in your practice area. Now, the second point assumes that your corporate field of practice is very close to the practice of your future business. If you are too far afield, you may work at cross purposes to your goals, so it pays to be mindful of the area you are working in and will be working in. The author will explain her career path and outline how capital project work and safety are closely associated, and how doing both has actually aided her business.

In the next item, if you are lucky enough to travel the world, work with many different kinds of people, work with many different contractors, and work with vendors and other owners and academics, you are building your network. All of these people become potential clients as you set up your own business. It pays to know a lot of different people, and more importantly, for them to view you as competent and knowledgeable in your field. If they cannot hire you, they may recommend you to others. Some of this work has happened to me, and it is very gratifying.

Lastly, a corporate career with a good company will allow you to obtain and to maintain your key credentials, namely, a professional engineering license and other certifications. This "sets you up" for your future very well, and you do not have to spend time and resources as you are trying to get your business off the ground.

In summary, Deborah Grubbe will outline in her presentation four ways that a corporate career will assist your transition to an entrepreneur. Deborah will cover the specific attributes of where and how a corporate role prepares you for entrepreneurship. Additionally, she will share some key professional elements to consider when preparing yourself for that career step.

(65b) The Reality is that Noble is a Wonderful Business Concept but Not a Sustainable Product

Diana Matonis, Midstream Energy Group, Sugar Land, TX

Starting a small business is for many an irresistible challenge. Many think that the first thing needed is picking a Name for the new company. At the kick off meeting, exuberance and ideas are exchanged but not one presented Name quite “fits”. A pragmatist in the group tells the charged up group that maybe we can't agree on a Company Name is because we don't have a product or market defined yet. It might be best to wait on the name that will come once a product or even market is agreed to. Someone stands up and in all seriousness declares, “Yes we have a business! The core theme of our business will be the usage of innovative technologies in order to have a positive impact on our world.” Sounds exciting, fulfilling, rewarding, doesn't it? They all agree this is it! The pragmatist obviously sees the glass as half empty, not half full. They all sit admiring themselves as they now have a mission and vision statement all figured out. The reality after this is that financing is needed, which means a product and a sound business plan. The excited entrepreneurs go investor to investor; traversing the financial world to find only a few that will give them a couple of their minutes of their time.

Anyone that has been through this and actually succeeded later will have compassion for your current quest and genuinely listen. After your excited elevator pitch the successful business person, entrepreneur, will in all sincerity and concern inform you the idea is a bit too broad although definitely a noble idea but will ask you to comeback when you have a more defined business plan and product. This is the place most new businesses come to a grinding halt as discouragement and reality set in. Many will go back and continue their day jobs discouraged they cannot be entrepreneurs. So what went wrong? What did they not do? This session will cover some of the cases and paths forward to starting a business, but will not give you the magic formula to assure your success in starting one.

Extended Abstract: Starting an IT Business

By Jerry O'Brien, President, Process Data Control Corp

As both presidential candidates told us repeatedly during the recent elections, small businesses are the economic powerhouse that drives job creation and economic growth in this country. In the case of Information Technology (IT) businesses, it is true that automating business processes that were formerly handled in less efficient ways can reduce the work force, however, such reductions are usually more than offset by employment increases as those who benefitted from efficient operations grow and prosper. This paper explores some of the general categories of IT-related products and services that entrepreneurs might consider, alone and in combination with each other, as plans are developed for their new business enterprises.

The growing importance of the so-called Knowledge Worker in many, if not most businesses today is creating enormous opportunities for new business start-ups that focus on IT products and services. Before investigating those opportunities, we discuss below a working definition of IT offerings within the "products and services" context.

What Are IT Products and Services?

The traditional definition of a product is something that a buyer of it can either walk out of a store carrying the item in a bag or box, or take the item from a deliveryman on the front porch. This quaint definition still applies when one is purchasing computers and components for them, but comes up short when looking at today's product menu. For example, what about the Apps I buy for my Smart Phone, or the latest version of QuickBooks? The delivery mechanism has certainly changed for buyers of software, but the basic concept of a product seems to hold.

But then there's the pesky field of SaaS (software-as a service) products, which deliver a specific benefit, much like a product, but require a continuing input of services, such as web hosting and software upgrades. Have we crossed the line into IT Services when considering such products? We may have stretched the concept a bit, certainly, but we are still looking at items purchased to accomplish a specific objective, which is more related to products than professional services; thus, even SaaS purchases may be considered products under this expanded definition.

IT Services comprise offerings that have traditionally been purchased either by the hour, or on a "lump sum" basis; and, unlike products, not much has changed in that regard. Services rely exclusively on the expertise and professional labor of the service provider. Though not a large factor now, we may see IT services that are furnished by robotic brains instead of human beings in the future; if so, the standard compensation model for purchasing services may need to be reconsidered.

Types of IT Products and Services?

Generally speaking, there are two categories in which IT-related products and services may be placed: (1) those that improve customer products and services in a fundamental way; and (2) those that improve the business model or capabilities of customers in such a way that they can provide their products and services more efficiently.

IT offerings that improve customer products and services will always thrive as long as they impart value to those products and services more cost-effectively than competing sources of similar technologies. For example, the latest advancements in new car design are in electronics, primarily to enable drivers and passengers to access information and services from a dashboard-mounted console. Only domain experts who are steeped in the technical knowledge and requirements of their future customers can hope to compete in this technology category. That fact limits the prospects of new IT businesses whose principals lack this type of background.

IT products and services that make business more efficient or effective require focused expertise, as well, but the domain is not necessarily related to the customer's area of interest. For example, numerous start-ups provide value to customers in the mobile computing field. The emergence of technologies that enhance targeted e-mail communications for public relations and marketing is another IT service area that falls under that category.

Business Process Improvement ("BPI") is broad fields in which companies that provide IT products and services are flourishing. BPI is defined in Wikipedia as "...a systematic approach to help an organization optimize its underlying processes to achieve more efficient results." BPI is different from the mobile computing and public relations examples in that it's goal is to make customers more effective businesses in all aspects of their operations. Although specialized knowledge in the BPI field is needed, no domain knowledge relative to customer markets is required of BPI vendors. In fact, the lack of such knowledge is often helpful in developing fresh insights and innovative solutions to problems that experienced workers have been unable to solve. New businesses that provide BPI products and services may have no specific domain knowledge other than in the field of BPI, however, older firms that have added BPI to their offerings often enter the field with domain expertise in an area that is heavily invested in procedures and data management, such as regulatory compliance.

Hallmarks of Successful BPI Products and Services

Because the value-added of IT products and services related to BPI is greater efficiency and effectiveness, it is of paramount importance that the design and operation of those deliverables be equally efficient and effective. In a similar vein, customers expect their vendors who furnish BPI products and services to operate efficiently because they have implemented BPI technologies themselves. Vendors who offer mobile computing platforms and electronic communications based public relations tools would have a difficult time convincing customers that their products and services impart value, if they themselves have not implemented such advanced technologies in their business operations.

The most important business planning concepts for IT businesses that provide BPI products and services are: (a) a sustainable business model; and (b) a compelling value proposition. Wikipedia defines "business model" to mean: "...the rationale of how an organization creates, delivers, and captures value." This rationale is all-important for new businesses, because the lack of focus on a clear mission is often cited as a leading reason why new businesses fail. BPI vendors need to adopt business models that enable them to explain optimal ways in which their customers can employ their products and services. Sustainable business models are self-propelled and adapt to meet changing needs.

Value propositions are clear and concise descriptions of why customers are better off with the products and services that the business offers, than without them. Compelling value propositions are virtually "stand alone" justifications for moving forward with offerings of the IT business; in other words, they are convincing with minimal salesmanship required. In today's marketing environment, where buying decisions are often made based on the materials furnished through websites and social media, compelling value propositions are "a must."

New IT businesses that offer products and services that can positively impact how their customers do business, and utilize a sustainable business model and compelling value proposition to position themselves in a competitive market, will be far more likely to succeed than those that do not. Starting such a business requires a commitment to these fundamental building blocks of a successful enterprise. If the enterprise is focused on selling a BPI product or service that can benefit the company selling it, then by all means do so and use the experience as an illustration of adding value from the deliverable.

After Selling BPI Products and Services

Improvement is not a static concept, and this is especially true in the BPI field. New businesses that deliver BPI products or services would be well-advised to view each sale as the beginning point in forming a customer relationship based on continual improvement of that company's business operations. This can be accomplished by using a variety of techniques. First, customer service is itself an opportunity to get to know customers and their needs better, while solving the issue or problem for which they contacted you. Even customer needs that cannot be met with your current products and services should be carefully documented when new information about them is learned. Such information may lead to future enhancements to products and services that will significantly impact their future viability in the marketplace.

Implementation is another area where constructive relationships may be forged. Many companies make the mistake of rushing through the implementation phase to avoid potential problems and ensure that customers are "in production" as quickly as possible. However, unless the BPI product or service is being deployed to address an emergency, which is rarely the case, the vendor should use implementation as a vehicle to expand their knowledge of and relationship with their customers. Even BPI vendors that provide novel products or services will soon confront pressures from competition if they are successful, because the Internet is very effective at "spreading the word" to both potential future customers and tomorrow's competitors alike. A positive implementation experience could be the deciding factor that influences your customer's buying decisions when competitive products emerge.

The Personal Challenges in Starting and Maintaining a Consulting Business

Jack Hipple

Innovation-TRIZ
Tampa, FL

www.innovation-triz.com

Starting a consulting business does not happen in a vacuum. Unless the consultant is a hermit, there are many concerns revolving around:

1. Cash Flow. Making contacts, meeting with and talking with potential clients will consume time in sales and marketing that produces no immediate income. It is possible and normal that there could be up to six months with no significant income. That means that a savings nest needs to be in place prior to starting a consulting business. The only exception to this may be a situation where a previous employer or associate has made a firm commitment, with a contract, to buy a certain level of services in advance

2. Financial Reviews. When a steady employment income disappears from a credit or loan application, a new consultant may find it more difficult or more expensive to borrow money. This needs to be considered in the planning to start a new business. Purchasing a new car or refinancing a house is best done while still employed.

3. Erratic Income. Even when income begins to be generated by a consulting business, it is likely that it will be highly erratic. The spikes may be very high, but there may be valleys that are zero for a short time. It is prudent not to go overboard in spending when the income spikes are high.

4. Family Concerns and Spouse Styles. Marriages are frequently made up of two people who complement each other. The person starting a new self-employed business may be a natural risk taker, while the spouse may be used to a check every week without fail. Discussions about these kinds of differences need to be openly discussed prior to starting the business. There may need to be a frank discussion about how long a period of time is tolerable prior to re-entering the private or public employment market.

A new consultant, especially one who has been terminated from a corporate position, will need moral support from their spouse and children. If your children are old enough to understand, explain the realities of today's world to them. Long term employment guarantees are a thing of the past. If some short term financial pullbacks are necessary, ask for their help.

5. Circumstances. People start consulting practices for any number of reasons. It could be retirement occupation. It could be based on the belief that the subject matter knowledge will be better compensated outside their current employer. It could be due to a simple downsizing or outplacement in a poor job market and a simple desire to maintain a given standard of living. Each of these circumstances will dictate to some extent how the startup of such a business is done.

6. Marketing. Individuals who have been technical experts in a corporate or engineering design firm may never have had to sell themselves and their expertise to anyone. It is also possible that no one outside their firm knows about this expertise. There are several simple ways to begin to develop inexpensive marketing outlets while the business is developing. These include professional societies such as AIChE and making presentations at conferences and regional workshops. Each major metropolitan area has a local AIChE section always looking for speakers. These cost almost nothing.

Setting up a web site and having a presence on LinkedIn™ is essential. These are very inexpensive things to do. Never throw away a business card and make contact with all past professional associates and make them aware that you have started a new business.

Identify professional society training opportunities. These not only pay, but introduce a new consultant to potential client at no additional expense.

Make your availability know to local and national technical publications and send articles for review.

7. Past Employer. If your situation has resulted from a "downsizing", remember that these go on all the time in today's world and have lost much of their stigma. Of course, there are always situations where these are in fact a result of significantly poor performance or some kind of unethical behavior, but assuming this is not the case, don't hesitate to maintain contact with your previous employer's associates you worked with, including your previous manager. It occasionally happens that they may be following you out the door and may need your help as well! If your skill was well recognized but not needed to the same degree, it may be possible to negotiate some contracting opportunities in the short or long term.

As we can see, many of these issues are not directly related to the daily cash flow of the business but have their roots in the personal and relationship side, including your own family. Make sure to take all of them into account when starting a consulting business.

Overview of P.E. Licensing Process in United States

William R. Parrish, P.E. and Cory D. Jensen, P.E.

Professional engineering licensure was established in the United States to protect the health, safety and welfare of the public from unscrupulous or incompetent decisions of engineers. Licensure provides a proven and accepted assessment of experience, education and ethical standards.

This presentation gives an overview of the benefits of being licensed, the key players involved in becoming licensed and the path to licensure. The overview presents the background information to help put the other session presentations in perspective. More details about licensure are available in the March issue of *Chemical Engineering Progress (CEP).*

Value of being licensed

Being licensed shows other that you have a strong commitment to high technical and ethical standards. It can provide you with better career opportunities and provide a competitive edge when seeking employment. Many employers state that they would choose a licensed engineer over an unlicensed engineer, given both are otherwise equally qualified. Also, companies often offer incentives to encourage their employees to become licensed.

Some engineers think that licensure is important only in consulting or expert testimony. However, licensure is often required in firms doing design, construction and startup of processes. State laws also may include those teaching "advanced" engineering subjects be licensed.

Some states allow an "industrial exemption" for large companies who have the financial resources to cover liabilities arising from employee decisions. Most companies encourage licensure but many oppose eliminating the industrial exemption if they have operations in multiple states because it requires their engineers to be licensed in multiple states. There is a push to eliminate the industrial exemption. This important topic is discussed in another presentation.

No one can predict the exact future course of their career. However, being licensed may provide unexpected rewards and opportunities.

Key players in licensure

Becoming a P.E. involves education, experience and examination. Besides your coworkers, who can help guide your job experience, there are two important players in becoming licensed. The first is your local engineering board that establishes the standards for education and experience and ability to take the examinations. You will work with the board throughout the licensure process.

The second important player is the National Council of Examiners for Engineering and Surveying (NCEES) (www.ncees.org) . This is a nonprofit organization whose members are the various engineering licensing boards. The NCEES was created to provide some uniformity of standards of education, experience and examination. The membership establishes Model Law to offer uniformity. However, Model Law is a set of suggested best practices and no board follows Model Law exactly. All exams are administered and graded by the NCEES.

The licensure process

Obtaining a P.E. involves the following steps:

1. Having the appropriate formal education
2. Passing the Fundamentals of Engineering (FE) exam
3. Obtaining the appropriate experience (engineering internship)
4. Providing required documentation to board
5. Passing the Chemical Professional Engineering (PE) exam

While the general requirements are uniform across all jurisdictions, exact details vary. It is critical to work with your local board so that you understand and meet the exact requirements. A brief synopsis of each step is given below. More details are provided in the *CEP* article and in other presentations in this session.

1. Formal education: All jurisdictions require BS engineering degree, or equivalent, from an EAC/ABET-accredited four-year engineering program. Foreign degrees and other degrees are considered but must be approved by the board.
   Current Model Law states that as of 2020, engineers taking the PE exam must have a Masters or equivalent (MOE). This requirement is strongly opposed by AIChE and many other organizations. More details are presented in another presentation in this session and in the *CEP* article.

2. FE Exam: This exam tests your fundamental knowledge of engineering. It is easiest to pass when you are first eligible to take it (usually senior year or shortly after graduation). Another presentation and the *CEP* article provide more details of the exam and the upcoming changes in the exam.

3. Practical experience: Your work experience must be appropriate engineering practice. Typically four years are required although some boards permit fewer. In some jurisdictions advanced degrees can reduce the number of years required.

4. Documentation: The exact format of documentation and what is required varies between jurisdictions so it critical to work with your local board on this. However, all will require proof of education, passing the FE exam and work experience. The work experience will be supported by several references, who will also vouch for your ethical behavior.

5. PE Exam: This is the final step in becoming a licensed professional engineer. The exam is designed to be passed by a "minimally competent engineer with four years of practical experience". This exam tests the practical application of your engineering knowledge. Both a session presentation and the *CEP* article provide more details on the exam.

Summary

The AIChE recommends that its members become licensed to help ensure a competent and ethical chemical engineering workforce. While the process may seem daunting at first, working with your fellow employees and your local board will make the effort much easier. Each year about 400 chemical engineers become P.E's. They have made a significant investment in themselves and possibly in their future career.

(78b) Licensure--An Academic Perspective

S. Ranil Wickramasinghe, Department of Chemical Engineering, University of Arkansas, Fayetteville, AR

Most Chemical Engineering faculty members are not licensed. Why? Is it important for faculty to be licensed? Most undergraduates take the Fundamentals of Engineering exam but do not take the Professional Engineering exam. Why? Should faculty be licensed? Is licensure important for faculty who are conducting research? Should we encourage faculty to obtain their PE license? In this presentation these questions will be discussed.

TESTING FOR COMPETENCY - THE PROFESSIONAL ENGINEERING EXAMINATION

Anne Bertelsmann, PhD, PE, Galveston Bay Refinery, Marathon Petroleum Company LP, Texas City, TX
And
Denise Chastain-Knight, PE, CFSE, CH2M HILL, Augusta, GA

Introduction

The decision to pursue engineering licensure is a personal one, and the road includes many hurdles to overcome. Engineers take this road for a variety of reasons, with different motivations and experiences along the way, and finish the trek with the PE Exam. The Principals and Practice of Engineering Examination (PE Exam) is a nationwide exam used to ensure that licensed PE's are "minimally competent". The exam must include test content applicable to the competency being measured, effectively differentiate between competent and not competent candidates, and provide a consistent competency measure from administration to administration. This paper will define minimal competency, discuss the purpose of the Professional Engineering Exam, exam format using the Chemical Engineering Exam as the example, development methodology, item quality metrics and offer suggestions for test taking preparation. It will also describe the function of the NCEES Exam Development Committee and their role in developing a fair, effective exam, and explain why it takes weeks to get a result for a computer scored exam.

Path to Licensure

Whether an engineer undertakes the path to licensure early or late in their career, each must successfully negotiate the same hurdles. Each step in the process has a specific goal to either develop or to measure the engineer's capabilities toward professional competency. Figure 1 on the next page illustrates the path to licensure.

The quest to licensure begins with obtaining a Chemical Engineering degree. The next step is to pass an exam that measures competency on engineering basics. This first exam is the Fundamentals of Engineering Exam (FE) – also sometimes called the EIT (Engineer in Training) Exam. The FE exam is typically taken either during the senior year in college or shortly after graduation. The types of items resemble "textbook" questions and are designed to test understanding of the engineering fundamentals as taught in a typical engineering degree program. Currently, the FE examination consists of a general morning session (common to all disciplines) and a choice of seven discipline specific afternoon modules. Starting in the beginning of 2014, the FE examination will convert to a computer based test, which is discipline specific (i.e. no general module). For more information on the FE exam content and the conversion to computer based testing, refer to the NCEES website (http://ncees.org).

Figure 1: Path to PE Licensure

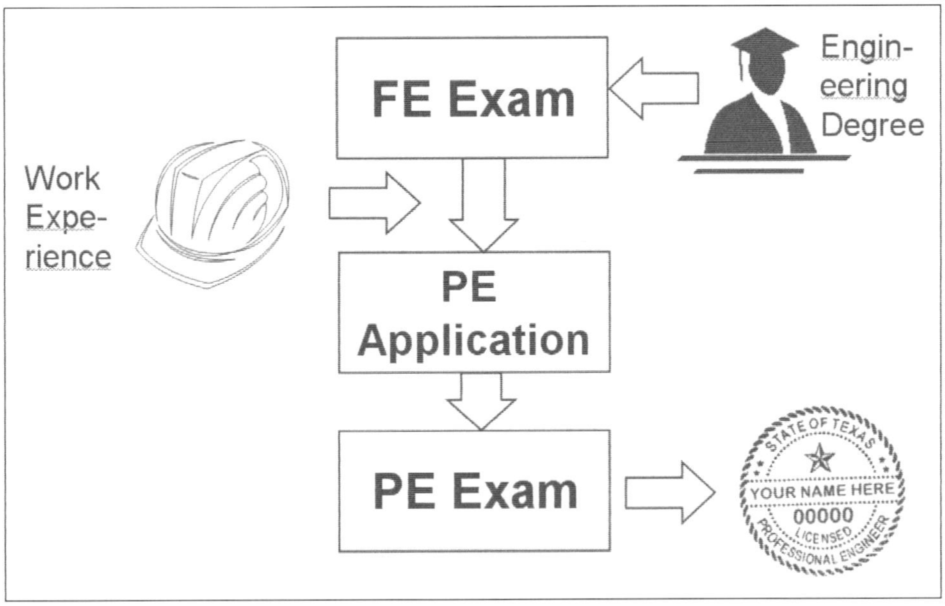

Following graduation, engineers enter the practice phase of their careers where fundamental knowledge is applied to real world application. In this phase of their careers, engineers use the theoretical principals learned from their education to solve problems that ultimately deliver physical products to market. Most states require a minimum of 4 years engineering related work experience before allowing candidates to sit for the PE exam. In order to become licensed, the candidate needs to submit an application to the state licensing board, describing education and experience, documenting passing of FE exam, and providing character and skills performance references. Format of the application and requirements for references, transcripts etc. vary from state to state. Once the application is accepted by the board, the candidate can register for the PE exam. After passing the PE examination, the engineer is licensed and can obtain the PE stamp. It is through this process that the state licensing boards ensure that only qualified candidates become licensed.

PE Exam Purpose

The PE exam is the final step in the road to licensure and its purpose is to ensure that only competent engineers become licensed. This helps to uphold the high standards of licensure and protects the health and welfare of the public. Since the exam is the same nationwide, it ensures consistency between states. Thus, one of the main requirements for a PE examination is that it can distinguish between competence and incompetence and that it can demonstrate this reliably and consistently. It is important to remember that "competent" should not be equated with "expert". The purpose of licensure is to identify individuals who have at least the minimum knowledge at which a person can practice professional engineering in such a manner that will safeguard life, health and property and promote the public welfare. That is why the term "minimally competent" is used.

A minimally competent engineer:

- should be able to perform identified tasks
- should meet minimum education and experience requirements
- is beginning to specialize in a sub-discipline
- may safely practice in another sub-discipline even though he/she does not practice in this area on a daily basis

The definition of "minimally competent" has implications on how the questions on the exam are written. The PE exam differs from the FE exam in that it addresses both the principles and the practice of engineering. So, while the FE questions are typically textbook related, questions on the PE exam are predominantly practice related. The questions build upon the experience that an engineer has gained during his or her first few years of working and need to recognize that chemical engineering experience can be gained in a variety of areas. The difficulty level of the questions is geared towards an engineer with about four years of experience.

PE Exam Content

The content of the examination is determined through a survey of practicing chemical engineers (PAKS study = Professional Activities, Knowledge and Skills), which is repeated about every 6-8 years to ensure that the examination stays current. The PAKS process includes conducting a technical content survey with input from practicing professionals. This survey is conducted over a period of weeks to ensure that the respondent pool size is adequate and is sufficiently diverse to be representative of the overall profession. Survey results are analyzed by a team of licensed Chemical Engineers, most with no prior exposure to PE exam development work, under the direction of a psychometrician who ensures that proper procedures are followed. The PAKS team will determine both the domain (subject area) content and item weight of the subdomains. PE exam content is established based on the PAKS study, and is published in the format of the PE exam specification available from the NCEES website (http://ncees.org/exams/pe-exam/) and is summarized in Table 1 on the next page.

The ChE exam specification was recently updated and became effective for the April 2011 exam administration. It includes six major domains and fifteen sub-domains that investigate theoretical and practical knowledge. Exam content is distributed within the major and subdomains according to the percentages shown in Table 1 to ensure a consistent content basis for every exam. The specification also includes guidelines for topical content within each sub-domain. For example the exam will include 4% of the items from sub-domain VI. C. Operation. Figure 2 includes example content from the 2011 exam specification.

Table 1: April 2011 ChE PE Exam Specification Summary

Knowledge Area	Approximate Percentage of Examination
I. Mass/Energy Balances and Thermodynamics	**23%**
A. Mass Balances	10%
B. Energy Balances and Thermodynamics	13%
II. Heat Transfer	**16%**
A. Mechanisms	9%
B. Applications	7%
III. Kinetics	**11%**
A. Reaction Parameters	6%
B. Reactors	5%
IV. Fluids	**16%**
A. Mechanical-Energy Balance	12%
B. Flow and Pressure Measurement Techniques	4%
V. Mass Transfer	**14%**
A. Phase Equilibria	5%
B. Continuous Vapor-Liquid Contactors	8%
C. Miscellaneous Mass Transfer Processes	1%
VI. Plant Design and Operation	**20%**
A. Economic Considerations	1%
B. Design	10%
C. Operation	4%
D. Safety, Health, and Environment	5%

Figure 2: Sample Exam Specification Sub-domain Content

C. Operation 4%

1. Process and equipment reliability (e.g., testing, preventive maintenance, startup/shutdown procedures, robustness)
2. Process improvement and troubleshooting (e.g., debottlenecking, experimental design and evaluation, optimization)

PE Exam Format:

The PE examination is given twice a year - once in April and once in October. It is an eight hour exam consisting of a four hour morning session and a four hour afternoon session with a lunch break in between. There are a total of 80 multiple choice questions (40 in the morning, 40 in the afternoon). Each question has four answer choices of which one and only one is the correct answer. Each question is stand-alone, i.e. it is independent of the other questions on the exam and can be solved with the information given in the problem, sometimes with the help of standard references. Some questions involve calculations while others are qualitative. All questions count equally and there is no penalty for guessing.

Each of the 80 multiple choice questions on the exam is of the same general format. In exam development, questions are referred to as "items" rather than questions because not every item has a question mark (like the one in the example below). Figure 3 illustrates the anatomy of an item, i.e. its general format and identifies the different parts of the item.

Figure 3: Anatomy of an Item

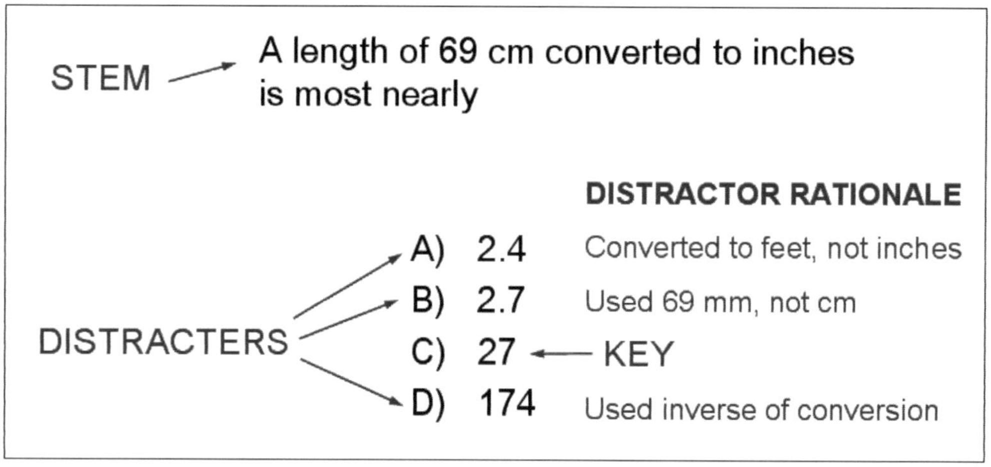

Each item consists of a stem and four answer choices. The stem (or problem statement) identifies what calculations are needed. It includes the information that is provided to solve the problem (sometimes even in table format) and states what is required for an answer. Beneath the stem are the four answer choices. Of these four answer choices, one (and only one) is the correct answer – the key. The other three are so-called distracters. They are incorrect answers – but they are not random numbers.

Each of the wrong choices represents a typical error that a less than minimally competent engineer (or someone not familiar with this specific topic) may make. Each distracter has to have a "rationale", i.e. an explanation which error was made. So, when a candidate is taking the exam and calculating a number that is exactly one of the given answer choices it does not necessarily mean that the problem was solved correctly. Maybe the candidate just made one of the "expected likely errors"...

The distracters can have a significant impact on the statistics of an item. If one of the distracters is obviously wrong (such as a negative absolute pressure), then even less than competent candidates will disregard it. That reduces the feasible choices, improves the "guessing" performance, and increases the chance that a not minimally competent candidate gets the question right. The Exam Development Committee strives to develop items that include the key, and three equally likely errors as distracters.

Exam Development

The PE Examination (as well as the FE, and the Surveying Exams) is a service that NCEES provides to the licensing boards. As such, NCEES provides the framework and the oversight for the exams, but the responsibility for the content rests with each exam committee. The committee writes the items, assembles the exams and reviews the statistics. NCEES has established rules and procedures that the exam committees follow to ensure consistency. Committees report back to NCEES with their exam quality measures. NCEES also provides the logistical support (psychometric consultant, publishing and printing, administrative support, maintaining records and the item bank, administering the exam, scoring, etc) and provides all security functions. For the larger exams, NCEES also pays and arranges for volunteer travel. For the smaller exams, this is the responsibility of the sponsoring society.

The Chemical Exam Committee is the group that actually writes and reviews the Chemical PE Exam. To be on the committee, one must be a licensed PE and express interest through the NCEES website or through a committee member. The committee has about 50 active members and meets twice a year in Clemson, South Carolina at the NCEES headquarters. Not everyone comes to every meeting, of course, so typically 30 – 35 committee members present at any given meeting. During the meeting items are written and reviewed by committee members who work with one another to critique and improve the items. Task teams also work to assemble, review and pretest exams.

Before an exam is administered, each item will have no less than six documented formal reviews and likely have been assessed by more than 20 people. Committee members are asked to complete detail oriented and often challenging engineering work and can earn PDH (professional development hours) for this activity. Developing the PE exam helps committee members to keep their engineering skills current and to give back to the profession. Without the work of this committee, the PE exam and with it the licensing wouldn't happen.

Preparing an exam that effectively and consistently measures minimal competency is a challenging task that succeeds or fails based on the ability of the committee to predict item performance. Each committee relies on a structured process and employees a variety of quality tools and techniques to improve the opportunity for success. Figure 4 is a workflow of the exam development activities. Items are written by a PE and submitted for review. Many items are prepared by the committee members at each meeting, while others are submitted by PEs who do not attend the meetings. The individuals who write and review items are referred to as Subject Matter Experts (SME). The author and two reviewers will independently work the item and develop the rationale for each distracter. Through this process the item is often revised and edited to improve its presentation and quality. This step is considered one of the most critical activities of the committee because items are the building blocks of the exam and each must meet high quality standards. Each committee member is responsible for item quality and is granted authority to modify draft items as needed to meet quality objectives, or reject items that fail criteria and cannot be edited to usable items. Finally the item is given a final check by the committee chair (or authorized designee) to verify that documentation is complete, both reviewers agree to the item content, and all quality checks have been completed, and approves the item for acceptance into the item bank. The item bank is a repository for items that are ready to be placed on exams.

Figure 4: Exam Development Process

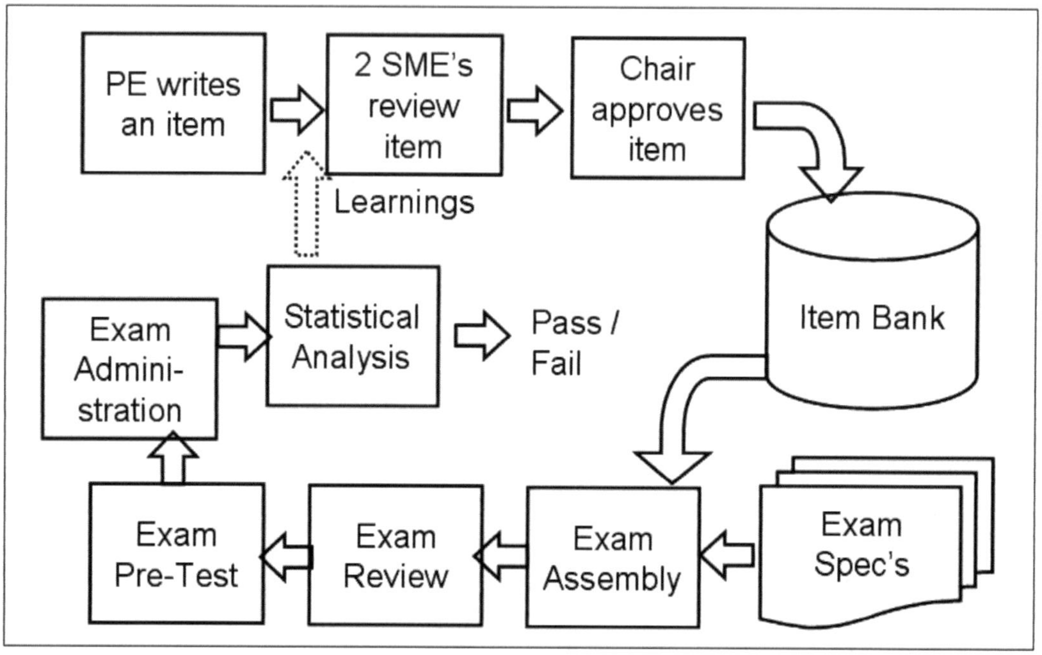

Exams are built, reviewed, revised, reviewed, quality tested and reviewed over a series of meetings. Each exam will have three independent team reviews, a pretest and a quality assessment. There are specific checks required at each workflow gate with task teams assigned to maximize the number of eyes that evaluate any given exam. An exam is compiled using items from the bank following the content distribution of the specification and sequenced in the specification order. The build team will then conduct a complete review of the exam, looking at each exam item for individual quality and checking for overall content flow and replacing items if necessary. The exam is then reviewed in two more meetings by independent teams who provide a "fresh set of eyes" and continue to improve the content to ensure that each item on the exam and the overall exam meet quality criteria. They also verify that the documentation is complete in accordance with the NCEES exam development requirements. Each exam must also have a pre-test where a group of committee members take the exam in as if they were actual test takers. After working all the items on the exam, they answer a series of questions about the items designed to predict item performance, then perform a distracter evaluation. Comments from the pre-testers are incorporated into the exam. In the final meeting before administration, a quality team conducts one last review of the exam verifying that all preceding checks (and associated documentation) are complete, all comments have been addressed, specification distribution is correct, and that author diversity limits and other NCEES criteria are met.

When the exam is administered and scored, data is compiled that can be used to evaluate the quality of the exam as a whole, and individual items. The data is analyzed and used to help the committee to learn what makes items perform effectively, and improve our performance prediction skills. Authors are provided a 'report card' so they know how items they wrote performed and have the opportunity to see how those items evolved in the review process. At each meeting, the committee will review items that had great statistics and try to find out what made them perform well. Similarly, we will review items that didn't perform well and identify traits that are undesirable in future items.

Item Statistics

Item quality is ultimately measured by the statistics generated from the candidate responses after an exam administration. Before discussing the statistical analysis of the examinee responses, one must first consider what makes an item a "good" item. The definition of a good item is one that is capable of distinguishing between "minimally competent" and not "minimally competent" candidates, does not discriminate against any particular group of test takers, and performs consistently in multiple exam administrations. To achieve those goals, it must be relevant, not confusing, and measure competency.

Most importantly, the content of the item must be relevant. Relevance is defined as the appropriateness of the topic and the difficulty of the item. The topic of the item must test knowledge or skills that are directly related to the exam specifications. For example: a question on high voltage power distribution may be relevant for an electrical engineer, but will not provide information on engineering competency of a *chemical* engineering candidate – no matter how well written it is. If the item does not relate to the exam specifications, it does not belong on the exam. Second, the difficulty level and complexity of the item must be such that a "minimally competent" engineer with 4 years experience can solve it in about 6 minutes. When writing items, authors often think about the interesting and challenging problems that are encountered in practice; however, if an engineer with 20 years experience requires several hours or days to solve the problem, what are the chances that a junior engineer will get the answer in 6 minutes? These type of problems, as interesting as they may be, do not make good exam items.

Another important principle to consider is that items should not be confusing to people who know the answer. When an item is confusing, a competent engineer who understands the concept being tested may not select the key, thus the question does not help to distinguish between competent and not competent. The so called "trick" question does not evaluate a candidate's competence and is considered a "bad" item to be weeded out in the exam development process.

It is equally undesirable to use items that give the answer away to people who don't know the answer. If the item includes clues to the key, then an incompetent engineer can choose the right answer without understanding the subject matter; thus the item does not measure competency. Clues can take many different forms:

- If some of the answer choices can be ruled out as impossible (like negative absolute pressures), the chance of guessing the right answer goes up.
- Grammar can give clues when one of the answer choices "stands out", i.e. is more detailed or does / does not match the grammatical structure of the item statement

For example, an item that asks about the 19th president of the United States, where three of the four answer choices are well known pop stars, does not require the candidate to know anything about US history to pick the correct answer.

Following an exam administration, a psychometric analysis is performed (psycho – metric = measuring the mind) on examinee responses. The analysis is solely based on information provided on the marked answer sheets without regard to the actual content of each item, but quite a bit of information about an item can be observed just from examinee responses. The psychometric analysis will evaluate a number of things. Of course, the simplest analysis is to determine how many candidates selected the key. Statistically, there is a 25% chance of guessing the correct answer because there are four choices. An item where less than 25% of the candidates selected the key may be flawed because it may have directed candidates towards preferentially selecting a distracter over the key. The next question is to ask *who* got the question right. Ideally, the candidates who got the question right were the same people who performed well on the exam overall. The analysis looks at the distribution of answers between the distracters and who selected them. This analysis also considers if all distracters were selected equally or if there was one distracter that was particularly attractive; maybe it even got selected more frequently than the key.

The objective of this analysis is to confirm the validity of each item as a tool capable of measuring engineering competence. An item that follows the basic three criteria (relevance, no confusion and no clues) can be expected to distinguish between competence and incompetence; however, sometimes clues, confusion or complexity can creep into an item unintentionally and be missed during the review before the exam is given. Statistics are then used to identify potential flaws. One such flaw is a mis-key. Mis-key is when the committee identified for example option A as the correct right answer, but option B is actually correct. Another flaw is an ambiguous item where more than one of the options could be correct. Despite the committee's efforts, such flawed items occasionally make it through the review process and are placed on an exam. The rigor of review process prior to exam administration minimizes the likelihood of a flawed item; however, it cannot totally eliminate them. The statistics after the exam will reflect item imperfections and provide an opportunity to compensate for item flaws in exam scoring so that candidates are not negatively impacted by flawed items.

Figure 5 is an example of the statistical analysis report for an item prepared by the psychometric consultant. One such sheet is prepared for each of the 80 items on an exam. The top left corner shows the "number of cases", i.e. the number of examinees who took the exam in this administration. In this example, there were 2910 candidates. So, this example is not from any Chemical PE exam where typically about 300 candidates sit per exam. The number below the case number (highlighted in yellow) is the Mean (or the P+ value). This number shows what percentage of the candidates selected the key; in this case, 65%. P+ alone indicates that the question is of appropriate difficulty for the average test taker; not too easy and not too hard. Statistically, the higher the P+, the easier the item.

Figure 5: Item Statistics

Number of Cases	2910
Mean (P+)	0.65
Delta	11.44
Pearson	0.33
Biserial	0.43
Mean of Passers	0.78
Mean of Failers	0.53
Mean Near Cut	0.68
Phi	0.26
DIF Grade	

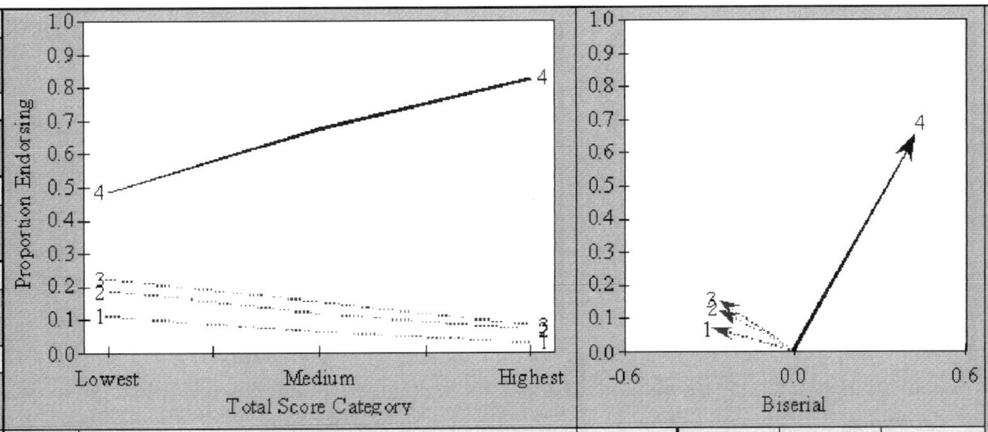

Alt.	Lowest	Low	Medium	High	Highest	%	N	Mean Scr	Biserial
1	0.11		0.06		0.03	6.7%	195	43.0	-0.29
2	0.19		0.12		0.07	12.5%	365	44.1	-0.27
3	0.22		0.15		0.08	15.4%	449	44.3	-0.27
4*	0.48		0.67		0.83	65.2%	1896	50.1	0.43
OM	0.00		0.00		0.00	0.2%	5	42.4	-0.20
Mn Scr	38.68		48.50		58.27				
Grp N	1025		1022		863				

The statistics not only evaluate how many candidates selected the key, but how selecting the key on this item correlates to overall exam scores. Test takers are divided into three groups of roughly equal size based on overall score on the exam; the low scoring third, the medium scoring third and the high scoring third. For each of these groups, the report in the table under the left graph shows how many selected which answer choice. For example on this question, "4" (or "D") was the key as indicated by the asterisk. Of the low scoring third, only 48% got the right answer, while 83% of the high performers selected the key. Thus, someone who performs well on the exam overall (which is an indication of competence) is significantly more likely to select the key than someone who scored low (may indicate lack of minimal competence). This spread shows that the question is capable of testing competency.

Figure 5 also illustrates that the incorrect answers were relatively evenly distributed between the distracters, indicating that none of the distracters appears to be overly attractive or easily ruled out. Another way of looking at the same data is presented on the right of the page. The table shows average overall scores for the people who selected each of the answer choices. In this case, the people who selected the key had an average overall score of 50, while the people who selected the distracters had overall scores in the range of 43 to 44. Five candidates with an average score of 42 did not answer the item. This clearly demonstrates that the people who selected the key were more likely to do well on the exam overall, and indicates that the item can effectively measure competency.

The data is evaluated in yet another way by the cumulative statistics in the lower left: Mean of Passers indicates that of those candidates who passed the exam, 78% selected the key. Mean of Failers shows that only 53% of the people who failed the exam selected the key. The larger the spread between these two numbers, the better the item is at measuring competency.

A handy way of summarizing the "who got the question right" analysis is the R-Biserial number. Basically, this number is related to the slope of the lines in the graph on the left. For a good item, this slope for the key should be positive and the larger the better because it expresses the difference in performance of the low and the high scorers. Theoretical range for the R-Biserial number is from -1 to +1, positive values indicate that the candidates who selected the key also scored higher on the overall exam. A negative R-Biserial number indicates that higher scoring candidates selected one of the distracters, which could be an indication of a mis-key or ambiguous item. The R-Biserial number is highlighted in yellow on the left and shown in the picture on the right. The R-Biserial number is important because it evaluates how well an item meets the three basic criteria. We previously observed that the sample item is of appropriate difficulty based on the percentage of correct answers (P+ value). The R-Biserial number shows that we did not confuse the competent candidates (a high percentage of high performers got it right) and apparently it did not give any clues to the less competent candidates (. It also appears to be relevant with regard to the topic since the high performers (those with better understanding of chemical engineering) did better on this than the low performers.

To further illustrate the concept of the R-Biserial number, consider an example of an item that might produce undesirable statistics. The item is asked on the final exam for a graduating class of chemical engineers: "Which bar in town has the most beers on tap?" One would expect a certain percentage of students to select the correct answer, but are those candidates the most knowledgeable students who spend a lot of time in the lab, or the "party crowd"? This question would likely have a negative R-Biserial number, indicating the overall least knowledgeable students are more likely to select the key, which means this item is not an effective measure of a graduating chemical engineer competency, nor is it relevant to chemical engineering knowledge and skills.

Determining Competency

The individual item statistics are the building blocks for determining the overall exam performance. To make the move from the individual item statistics to determining competency of the candidate, a passing score, also called cut score, needs to be established. Figure 6 on the next page illustrates the philosophy behind relating the score distribution to the cut point, or passing score. Total score is simply the number of items answered correctly. For the Chemical Engineering PE Exam, the total score has a theoretical range of 0 to 80. Most candidates have at least some knowledge and get lucky guessing on some questions, so scores on the very low end are uncommon.

If it were known upfront which candidates are competent and which ones are not, the scores of the two groups could be plotted as shown in Figure 6. The at least minimally competent engineers have higher scores, the not minimally competent engineers have lower scores. However, there will be some overlap between the two groups. The better individual items differentiate between competency and incompetency, the smaller the overlap. However, because there is a range of competency and because a person's performance can vary, there will always be some overlap and thus the cut point decision is not clear cut.

Figure 6: Score Distribution

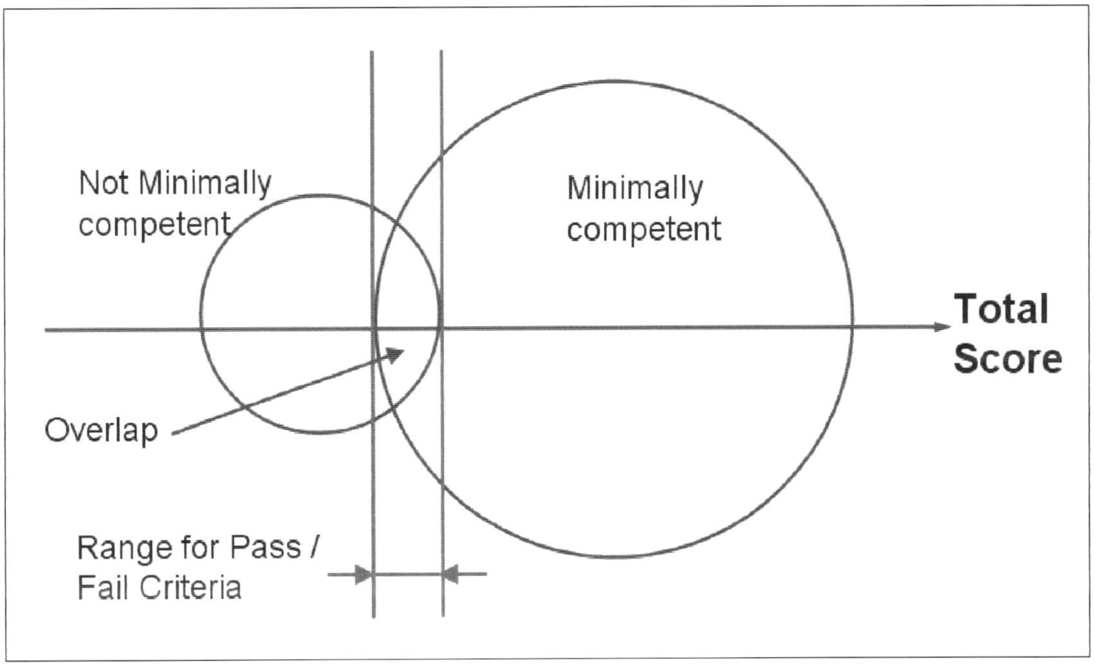

The overlap represents the range where the pass/fail decision should be made. All the at least minimally competent candidates on the right side should pass, all the not minimally competent candidates on the left should fail. With any decision point within that overlap range, there is a potential for competent engineers to fail and/or for less than minimally competent engineers to pass. Failing a competent engineer means that they will need to re-take the exam and includes the risk that they may give up on licensure. Passing a less than minimally competent engineer may allow an under-qualified person into public practice. Since the second type of error is more consequential, there is a tendency towards raising the bar when determining the pass/fail criteria.

Without knowing upfront which candidates are "minimally competent' and which ones are not (after all, it is the purpose of this exam to establish this), statistical measure are used to determine the pass/fail criteria, i.e. establish the "cut score". The cut score is the number of correct answers that a candidate needs to have in order to be considered "minimally competent". One way of determining the cut score is through panel of licensed PE's. This process is fairly involved, since it requires about 15-20 PE's who can volunteer 2-3 days of their time and travel to attend to cut score session. The panel must have good diversity (age, gender, area of practice, geographic location, demographics, etc…) and the cut score study must be completed before the results of the exam can be reported back to the state boards and ultimately to the candidates who are eagerly waiting for the scores. For these reasons, a cut score panel is usually only done when there was a change in exam specifications where the exam results cannot be directly compared to recent prior exams. The first exam where the format or specification is changed is called the "anchor exam" and cut point is established by a formal study.

For a cut score study the diverse panel of volunteers (all licensed PE's) meet and spend about a day to work the exam in a timed environment just like the actual administration. They can use any references, cannot consult with one another and they don't have access to the exam solution. The next day, they receive their score, and review and discuss each item as a group. They look at the statistics similar to those described above, but lacking the pass/fail metrics, review the solution, the distracters and discuss the topic and difficulty of each item. On the basis of that discussion, the panel members make a judgment call on how many "minimally competent" engineers out of 100 would answer this question correctly. A psychometrician guides the session and collects these estimates from each panel member. This data is then used in statistical analysis, generating a proposed passing score. The proposed passing score can then be adjusted (for example, to err on the "safe side"). This then becomes the basis for the pass/fail decision for the anchor exam.

Such a cut score study is not practical for every exam at every administration. So as long as there is no change in the exam specification, it is possible to "equate" the current exam to a previous exam. The equating process, heavily supported by a psychometrician, compares examinee performance on the current exam with examinee performance on previous exams. From this comparison it can be determined if the current exam is harder or easier than the previous exams, and cut point adjusted accordingly. If the data indicates that the exam is easier, the cut score is adjusted upward and a candidate will need to answer more questions correctly in order to pass. Conversely, candidates are not penalized if they take a harder exam because the cut point is revised downward and fewer questions must be answered correctly to pass. Equating is NOT the same as grading on a curve. When grading on a curve, the passing percentage would be fixed and the cut score adjusted to achieve that percentage. With equating, the competency of a candidate is measured on personal performance, and not dependent on the individual candidate's position within the test taking population, nor on the difficulty of the exam.

The psychometricians calculate other measures from the data. Coefficient Alpha measures the internal consistency or statistical homogeneity of the data, thus basically expresses the reliability of the exam. A measure called "Subkoviak" provides an estimate of the consistency of the pass/fail decision or an estimate of the probability that an individual would receive the same pass/fail decision on multiple administrations of the exam. All these measures are regularly reviewed to ensure that the exams can consistently measure competency.

After an exam is administered and all the answer sheets are scored, the statistical item analysis is performed. Exam Committees need to review potentially flawed items in case the key needs to be adjusted and the exam rescored. Then the cut point is determined, either through a cut score study with a panel or through equating. Only then can the pass / fail decision be made. State boards are notified with the results, perform state specific pass/fail assessments and then notify candidates of the results. NCEES recommends that only pass/fail should be reported to the candidates, though some states provide the actual score (or a "scaled percentage" based on a fixed passing percentage) to candidates. Candidates who fail get a diagnostics report so that they know how they performed in the various subject areas. NCEES recommends to only report the pass/fail decision for two reasons: The cut score can vary from exam to exam, so a candidate may pass one year and another candidate taking the exam another year may fail with the same score. The other reason is that the exam is designed to make a pass/fail decision on minimum competency. In that regard, the PE exam is different from an exam like the SAT, which is designed to measure "aptitude" over a wide range of abilities. The PE exam is not designed to measure ability accurately at the far ends of the scale. The PE exam does not distinguish between the "outstandingly competent" and the merely "extraordinarily competent engineer". Similarly, it cannot distinguish between the "totally incompetent" and the "absolutely clueless". Thus, a candidate who received a score of 75 out of 80 is not necessarily a better engineer than a candidate with a score of 72 out of 80 – both are "minimally competent" and pass.

The skeptical reader asks; "Does this really work?" Can an exam such as this really measure competency as the statistical "number magic" appears to indicate? The answer is found in the historical exam pass rates. NCEES does not report an overall pass rate for the exam, but breaks it up into a pass rate for first time takers and for repeat takers. Table 2 includes pass rates for exams given under the current Chemical PE Specification. The most recent pass rates for other exams can be found on the NCEES website.

One specific item to note about the data in Table 2 is that there is a gap between the pass rate for first time takers and for repeat takers. The pass rate for first time takers is typically rather high; in the 70 – 80% range for the Chemical PE exam. This high pass rate is because the examinee pool is already a pre-selected group. In order to be admitted to the exam, a candidate had to pass the FE exam, had to prepare an application with documentation of engineering experience, and had to be approved by the state board, a process that weeds out some of the less qualified potential candidates. Also, especially in chemical engineering where there are many jobs that come under the industrial exemption and do not require licensure, the people who decide to pursue licensure are already a non-random subset of all engineering graduates. This contributes to the relatively high pass rates for first time takers. The portion of the candidate population that passes on a repeat attempt includes in candidates that are minimally competent but may have been inadequately prepare in their first exam sitting.

Table 2: Chemical PE Exam Pass Rates

Chemical PE Exam Pass Rates	April 2011	October 2011	April 2012	October 2012
1st Time Pass Rate	78%	75%	74%	71%
Repeat Pass Rate	31%	29%	31%	20%

How to Prepare for the Exam

All the fascinating information presented in this paper aside, chances are the reader really wants to know how to prepare for the exam. Preparation philosophy will differ for a young candidate near the 4 year experience mark from that for a veteran engineer. One theory is that any minimally competent engineer can pass the examination without any preparation if he or she had unlimited time - time to look up references, read up on concepts and topics outside one's familiar area of practice, find the right equations and conversion factors, etc. But there is a time limit, and 6 minutes per question is not a lot of time.

Exam preparation should focus on building speed and efficiency for solving items. Through study and preparation, a candidate should become familiar with reference content, placing tabs to help find frequently used tables or compile a personal reference of frequently used information. Common equations and frequently used conversion factors should be memorized. Candidates should reactivate the habit of using math: solving equations, working with variables and using a scientific calculator instead of a computer. Speaking of calculators, it is strongly recommended to prepare for the exam using one of the approved models. Precious time can be wasted searching for the logarithm button on an unfamiliar calculator. Approved calculator models can be found on the NCEES website.

Exam readiness is achieved through lots and lots of practice. Preparation classes are useful for refreshing material basics, but are no substitute for rehearsal. A candidate should prepare by solving as many sample problems as one can find. By solving many, many problems, one will automatically remember the frequently used equations and conversion factors, enhance math skills and remember where to find information. The process also develops test day confidence because the items encountered on the exam have been encountered in similar form during exam preparation. Candidates should conduct one or more mock exam day(s) to practice pacing and practice maintaining focus for 8 hours. NCEES publishes a sample exam for each specification that has been developed by the exam committees in the same manner as a "real" exam. One author of this paper frequently recommends to potential exam candidates that they practice with readily available materials, complete a timed trial with the NCEES exam, then continue practicing with more materials.

Finally each candidate should prepare for the test environment and stage to be at peak performance on test day. Exams are administered in finite locations within a state and often require candidates to travel. Plan a travel itinerary that will allow arrival at the test site early, rested and unstressed. Have a satisfying but not heavy breakfast and make a plan for a relaxing lunch. Consider the test room conditions, dress comfortably and in layers. Be aware of the materials that are allowed in the test hall, and prepare accordingly. Candidates will be turned away without a valid government photo ID. Cell phones and other forbidden materials may be confiscated or admission refused. The exam is worked in a printed booklet, which contains sufficient space for working the questions and pencils are provided. Candidates need to bring an approved calculator, unapproved calculators will be rejected. The exam is open book, so candidates also need to bring any references that they want to use during the exam. References must be bound (no loose paper), but can be marked with flags. Review the test day policies on the NCEES website and plan accordingly (http://ncees.org/exams/exam-day-policies/).

Closing Comments

Professional Engineer Licensure is a personal decision and requires one to undertake a dedicated and challenging journey. Candidate should embark on this journey knowing that the exam development process is designed to ensure each exam measures for minimal competency in a consistent manner, and items are statistically validated. Finally, the factors influencing exam score processing timelines are understood. The authors of this paper encourage engineers to pursue licensure. We speak for ourselves and our fellow PEs when we say that nobody ever regrets the resources and effort spent obtaining a PE, and we appreciate the industry approval and instant expert acceptance granted by those two little letters following our names. P.E. Professional Engineer.

(78d) The Current Hot Button Issues in Domestic Professional Licensing

Amos Holt, Southwest Research Institute, San Antonio, TX

Amos Holt, a past president of ASME and current VP for Environmental, Safety and Quality Systems will review the issues currently receiving the greatest attention by the PE community which includes all of the various state and territorial licensing boards, the National Council for Examiners of Engineers and Surveyors as well as the numerous professional society stakeholders and educational organizations and societies. The discussuion will touch on the Model Law and its currect sections on the Masters or Equivalent (MOE) requirements as well as Industrial Exemptions. The various positions of key players covered including the position of a consortium of professional societies called "Licensing that Works" or LTW will be summarized. This group includes both ASME and AIChE and represents about 240,000 engineers thtroughout the world. Amos will tell you what you will likely be hearing more about in the next few years and how it might affect both current PE licensees and prospective PE licensees.

The California Divide – And Why You Need To Engage Your Legislators

Diane Spencer, PE, Lawrence Livermore National Laboratory, CA
and Emmett Miller, PE, consultant and AIChE Fellow, CA

Abstract

California is crippled by a great divide that has nothing to do with tectonic plates. California's professional engineering statute gives civil engineers free reign, allows electrical and mechanical engineers to work in their disciplines so long as that does not conflict with civil engineering, and acknowledges by title only that other engineering disciplines exist.

This division presents problems for engineers whether they have a professional engineering license or not. For example, Spill Prevention Control and Countermeasure (SPCC) plans are mandated by federal regulations. Such plans are prepared by chemical engineers all over the country, but NOT in California. SPCC plans are part of the exclusive domain of civil engineers.

A fire protection engineer was once told he could not design a fire suppression system because that was the exclusive domain of mechanical engineers. A chemical engineer was once told he could not submit a Risk Management Plan (RMP) (required by Cal EPA) without a civil engineer stamp because even the hazard analysis on a facility that uses chemicals is the civil engineer's domain. These are just a few examples where engineers are not allowed to work on projects that similarly trained engineers are expected to do in the rest of the country and the rest of the world.

So how do engineers in California navigate this regulatory nightmare? And how do we influence the legislators, engineering board, and others to nudge the statute in the right direction?

California's answer included forming a coalition of professional engineering societies that work together to present a united voice. The California Legislative Council of Professional Engineers (CLCPE) not only represents the interests of engineering professions to the legislature, but serves as a pro bono resource for the legislature on technical issues as well.

This paper will discuss: (1) our lessons learned from working with legislators; (2) California's professional engineering statute and why it is written as it is; and (3) why one needs to be engaged in the legislative process. Professional engineers are entrusted with protecting the health, safety, and welfare of the public. Getting engaged has the benefit of letting legislators understand the impact of their decisions and allows engineers to serve the communities where they live, and provide a safer, better community for everyone.

California's Problem

While most other states utilize some form of a generic registration model for engineers, California's model is discipline-specific, with civil engineering held on a pedestal. The definition of "civil engineering" in the Professional Engineers Act (California Business and Professions Code Sections 6700-6799.) is so broad that it encompasses nearly the entire spectrum of engineering. This overly broad definition, coupled with a mandate that only civil engineers can do civil engineering, means there is little room for those in other engineering disciplines to legally work in their own professions. The law allows overlap from civil engineering into all other disciplines, but other disciplines cannot overlap into civil engineering. So, not only are chemical engineers impacted, the other engineering disciplines are impacted as well.

Section 6731 states the following, with parts underlined for emphasis:

> "Civil engineering embraces the following studies or activities in connection with fixed works for irrigation, drainage, waterpower, water supply, flood control, inland waterways, harbors, municipal improvements, railroads, highways, tunnels, airports and airways, purification of water, sewerage, refuse disposal, foundations, grading, framed and homogeneous structures, buildings, or bridges:
>
> (a) The economics of, the use and design of, materials of construction and the determination of their physical qualities.
>
> (b) The supervision of the construction of engineering structures.
>
> (c) The investigation of the laws, phenomena and forces of nature.
>
> (d) Appraisals or valuations.
>
> (e) The preparation or submission of designs, plans and specifications and engineering reports.
>
> (f) Coordination of the work of professional, technical, or special consultants.
>
> (g) Creation, preparation, or modification of electronic or computerized data in the performance of the activities described in subdivisions (a) through (f)."

For many years the conventional wisdom has been that the term "fixed works," as used in the statute, (see Section 6731 above) somehow limits the scope of civil engineering, leaving room for other engineers to practice their craft. In order to clarify where that boundary lies, Senator Mimi Walters asked the California Legislative Counsel Bureau for a legal opinion (2011). The findings were shocking. "Fixed works" includes not only the structures, but also all the attachments to the structure. So, if a mechanical engineer designs a pumping system or a chemical engineer designs a fractionating column, and that hardware and the hardware is secured to a foundation, those engineers have crossed into the exclusive domain of the civil engineer. Other engineers are similarly impacted.

When a reasonable person reviews this list of studies or activities, it becomes apparent that the list precludes even scientists from legally studying the phenomena and forces of nature. And in most jobs, engineers are expected to produce engineering reports or to coordinate the work of other professionals.

How Did California Get Into This PE Mess?

The perverse nature of the current Professional Engineers Act in California is a result of good intentions gone awry.

In March 1928, a dam northeast of Castaic, CA, broke and flooded the valley below, wiping out the town of Santa Paula and flooding other towns downstream. Hundreds of people died and millions of dollars' worth of property was damaged.

The Professional Engineers Act was then created to ensure that only qualified people (civil engineer PEs) could work on dams and other fixed works. The legislators made the law as inclusive as possible to make sure there were no loopholes. Later, the Legislature expanded the Act to include Chemical, Electrical, Mechanical, Petroleum, Industrial and Metallurgical Engineering as Title Act Disciplines. With the passage of time, the Board for Professional Engineers, Land Surveyors, and Geologists (BPELSG, aka "Engineering Board") was empowered by the Legislature to create Title Disciplines on its own authority, which it did, eventually increasing the count to 15 title protected disciplines (Title Act disciplines), These are defined by Board Rules (16CCR 400-476) and have minimal stature in law. Along the line, Mechanical and Electrical Engineering were converted by statute to practice protection (Practice Act discipline)[1].

This patch job to add mechanical engineering and electrical engineering to the Professional Engineers Act has been ineffective (per Legislative Counsel's review). Inclusion of Title Act disciplines constitutes an even lower tier where those engineers are not able to take responsible charge of work within their field of expertise. In summary, the field is far from level.

The problem with the Professional Engineers Act has been known for many years. The difference now is the Legislative Counsel's legal opinion that was issued in July 2011 (www.CLCPE.org). According to that opinion, companies that rely on engineering services are on notice that the only engineering discipline allowed to conduct engineering work (not just supervise) in California is civil engineering. If a project requires a chemical engineer, a dual-licensed engineer will be needed to meet the competency requirements of both a chemical engineer and the legal requirements of a civil engineer. A dual license is needed for the other engineering disciplines as well.

---

[1] Four of the title disciplines (Corrosion, Manufacturing, Quality and Safety) were dropped ~1999 for lack of an NCEES exam.

The Industrial Exemption

Common to most states, California has an Industrial Exemption that allows companies engaged in manufacturing to use unlicensed engineers in their businesses— except for civil engineering work. The justification for the Industrial Exemption has been that such employers are capable of and competent in evaluating their own engineering talent, that they do not need interference from the state in doing so, and that failure that will impact the public is subject to strict liability.

There are exemptions for people who perform some specific and narrow civil engineering services. Architects performing architecture and building contractors following building codes are examples. But no general exemption exists for engineers who are not civil engineers if they perform "civil engineering." The Industrial Exemption even states that it does not apply to civil engineering.

With the recent legal opinion, however, all engineering practice in connection with facilities that are geographically fixed is now regulated as civil engineering, and the Industrial Exemption is therefore meaningless.

Engaging the Legislature

In the late 1940's, engineering societies in California organized the California Legislative Council for Professional Engineers (CLCPE) in order to work together as a resource to assist the California legislature, support laws that benefit the public and profession, and help keep society members informed about legislation that might affect them. Application of the Professional Engineers Act was not a problem to engineers in the state so long as the Engineering Board and regulators maintained a liberal interpretation. Likewise, CLCPE did not intensify its effort to change the law until the literal interpretation ultimately reinforced by the Legislative Counsel ruling came into focus.

CLCPE's efforts to change the law began in the 1980s. Our members helped draft a bill and found a senator to carry it. In California, the legislator who carries the bill is the "author" and organizations that support the bill, such as CLCPE, are "sponsors" regardless of who actually drafts the bill. Drafting a bill and finding "authors" is a process CLCPE repeated about a dozen times.

Let's reframe the process we used in California in terms that may help others take advantage of our lessons learned.

1) Coalition. We had a consortium of engineering societies working together as a coalition. This provided more credibility in that we were able to speak with one voice instead of individual organizations promoting separate proposals. A coalition also allows a sponsoring organization such as CLCPE to represent a larger portion of a legislator's constituency. AIChE local section leaders met with a senator to promote the one bill that chemical engineers sponsored without going through CLCPE. The senator listened intently about the impact of current law, asked a few clarifying questions, and then asked about how many chemical engineers were in the state. She was not satisfied with our answer, stood up and thanked us for our time.

Creating a coalition of several engineering societies enabled us to afford a lobbyist. AIChE and local sections are a 501(c)(3) nonprofit organization, which restricts how much money can be spent on lobbying. CLCPE however, was organized as a 501(c)(6) nonprofit, which allows for lobbying, but since the societies and CLCPE are all volunteer organizations, we do not have an intense political presence, and that restricts campaigning for or donating to candidates or political parties. Unfortunately, the opposition we face in California has major campaign funding and they openly support candidates that vote as the opposition wants them to vote.

Another advantage of working within a coalition of engineering societies goes beyond promoting legislation. The conversations we had when we met allowed us to gain insight into the problems each of the different disciplines were facing and we heard personal testimony about those various issues. This perspective has been invaluable.

2) Hiring a lobbyist. Over its lifetime, CLCPE has successively employed three lobbyists who helped with efforts to change the Professional Engineers Act, in addition to supporting the overall mission. A good lobbyist helps focus the discussion and facilitate consensus on language for a given bill. They also know how to navigate through the legislature, explain the bill or other proposals to legislators, legislative staff, and to others, and to find legislators who will author a bill. Typically engineers have full-time jobs, and working a bill is very time-consuming. The first of CLCPE's lobbyists was an engineer who later became Executive Officer of the Engineering Board. The other two were professional lobbyists with legal backgrounds.

3) Language for the bill. Earlier this paper discussed drafting language for a bill and then entering that bill into the legislative process (i.e., finding an author). "Entering" is the key word here. In the late 1990s, our bill attracted attention from the Professional and Technical Consultants Association (PATCA). In our quest to expand the law to allow engineers to practice their own profession, we needed to be careful not to adversely impact scientists (e.g., chemists) and other technical professionals. After several meetings with PATCA, we were able to work out compatible language. Going forward on subsequent bills, we deliberately engaged with the chemists (PATCA) early in the process.

4) Bill's "author". Over the decades we have been pursuing change, our bills have been authored by different senators and assemblymen. In several cases, the legislator had been an engineer in his/her former career. One such senator, who had been an aeronautical engineer, had commented that it took multiple disciplines to design an aircraft. A legislator with a technical or science background readily grasped the issue we were dealing with. One of the first committees to hear a bill dealing with engineering (i.e., the bills we were promoting) was either the Business and

Professions Committee (B&PC) in the Senate or the B&PC in the Assembly. When seeking an author we often approached members of these committees as well. The intent was to get early support and move the bill through the legislative process.

5) *Finding support*. In addition to sponsors, it helps to find other organizations that will support a bill through (1) letters, (2) testimonies, and (3) lobbying efforts. We collected binders of letters from individuals, municipalities, regulators, and corporations. These included Chevron, the Chemical Industry Council of California, and Genentech. The letters provided a lot of anecdotes on how this perverse law affected people. The volume of the letters attracted the attention of legislators.

Once a bill is scheduled for a vote, testimonies at the hearing can be very compelling. The city of San Diego, California Fire Prevention Officers Association, a division of the California Fire Chief Association, along with various individual department fire chiefs and fire marshals, individuals impacted by current law, and others have testified at legislative hearings over the years. The Dean of Engineering at University of California—Davis testified on behalf of the University of California at the hearing in January 2012, stating that the state is educating engineers who cannot work in their own professions. The dean was shocked to find out that the legislators still would not vote to support changing the current law.[2]

The third level of support is when companies/ organizations involve their lobbyists. However, we found it very difficult to convince companies that had lobbyists in Sacramento to get engaged at this higher level.

6) *Other curve balls*. We experienced numerous unexpected challenges. One such curve ball happened in 1997, when chemical engineers tried independently (without CLCPE) to change the law. The Engineering Board required that we provide a Sunrise Package[3]. Essentially converting engineers from Title Act (defined in by BPELSG Board Rules) to Practice Act (defined in the Professional Engineers Act) is viewed as creating a new engineering discipline. We were given one week to prepare this. Fortunately, CLCPE had prepared something similar 10 to 15 years earlier, which we were able to mine and then create the required Sunrise Package. The Engineering Board still did not support our bill. Note: Keeping a good archive of material produced in supporting legislation could have unexpected value in the future. The key is being able to find the needed archived information when it is needed. Another advantage is that many of the same people remained engaged in CLCPE for long periods.

---

[2] Incidentally, the western region of ASCE sent their lobbyist to testify in opposition at the same hearing where the UCD dean testified.

[3] Sunrise is a process under which an occupation or profession wishing to receive state certification or licensure must propose the components of the legislation, along with cost and benefit estimates of the proposed regulation

What We Proposed and Why

Our approach to changing legislation evolved over the years. As mentioned earlier, chemical engineers made an attempt in 1997 to raise chemical engineering to a Practice Act (defined by statute). Our opposition protested that our attempt was part of a turf battle among the engineering disciplines. Legislators do not want to be drawn into controversies (e.g., infighting). Our next attempt involved CLCPE, with the intent to define all the Title Act disciplines as Practice Act disciplines. Because we had a coalition involving more than 15 engineering societies, we were able to combat this accusation. Still we did not succeed.

We also shifted our approach and proposed legislation that was closer in alignment with generic engineering. Specifically, we proposed that engineers should be allowed to do work based on their education, experience, or testing. In talking to legislators and to our lobbyist, we learned that we needed to approach any change in engineering statute with baby steps. Adopting generic licensure language (similar to other states or similar to Model Law proposed by the National Society of Professional Engineers (NSPE) was too much at one time. Because of the heavy influence by civil engineering companies in the state, we drafted bills that would continue all the advantages afforded to civil engineering. Unfortunately, our opposition would not negotiate.

In 2000, CLCPE had caught the attention of the legislature sufficient enough that they commissioned a study of the two-tier approach (Title Act/Practice Act) used in California. The study considered natural overlap between the professions, whether there was enough distinction between the engineering disciplines to support this unique regulatory structure, and whether the three Practice Act disciplines (civil, mechanical, and electrical engineering) posed more of a risk to the public (relative to the other engineering disciplines) and thus justified the existing two-tier structure. (See **www.CLCPE.org** for study by the Institute for Social Research.)

The study strongly supported changing the existing regulatory structure. The task of making this change fell to the Engineering Board. The bill it sponsored was opposed by American Consulting Engineers Council (ACEC) - California (a contractors' association) and Professional Engineers in California Government (PECG, a public employee union). Both were well-funded with strong ties to legislators — and the bill failed. [PECG is allowed to donate directly to candidates and its 13,000 members individually campaign for legislators. That has a big influence on those legislators who might support our position.]

The study marked a shift in how we approach changing legislation; CLCPE did not sponsor bills after the study. Instead, CLCPE and our individual engineering societies supported a succession of bills that were sponsored by the Engineering Board or by companies and organizations that use engineering services.

The disadvantage of this approach is that we had less influence on making amendments to the language in the bills. So while sponsoring our own bills had not been successful, this other approach allowed us to continue to pursue fixing the Professional Engineers Act.

Another challenge we had was finding companies and organizations that would commit their lobbyists. Many of these companies were very willing to prepare letters of support, but were not ready to involve their lobbyists.

End Results

This paper provides numerous lessons learned from decades of experience in trying to change California law to allow all engineers to practice in their own professions without finding a civil (typically) or other Practice Act engineer to take responsible charge of the work. Despite years of effort, we have not succeeded in changing the Professional Engineers Act. We are debating our next steps. It is our hope that the lessons we learned will be helpful to others who find the need to engage their state legislators.

(78f) Panel of Speakers

Joseph Cramer, Technical Programming, AIChE, New York City, NY and Jeffery Perl, Chicago Chem Consultants Corp, Chicago, IL

Chairs Cramer and Perl will preside at a Q&A panel of the session Speakers. Questions from the audience, chairs and other panelists will be answered during this 30 minute panel period.

A Leadership Cycle for Technical Excellence

Mr. John T. Perez, P.E.
Cognascents Consulting Group, Inc.
11777-A Katy Fwy Ste 438
Houston, TX 77079
john.perez@cognascents.com

Keywords: Leadership, culture, excellence, metrics, competitive sustainability.

Abstract

In the pursuit of competitive sustainability amidst dynamic environments, individuals and organizations spend significant time monitoring and modifying behaviors to improve their overall performances. Furthermore, many organizations have identified the role of leadership in establishing, sustaining, and changing cultures in order to encourage optimal behaviors. Finally, many individuals have received formal leadership training designed to achieve and maintain high performance cultures. Despite all of the above, technical excellence remains a "Holy Grail" for some entities.

The author posits that technical excellence remains elusive for some entities due to a lack of focus on and understanding of the real drivers of excellence. The author asserts that technical excellence is not a value that individuals and organizations can espouse in a personal or corporate mission statement. Technical excellence is a condition and result of an individual's or organization's core, leadership model, culture, and behavior set. In order to achieve technical excellence, individuals and organizations must (1) nurture and safeguard their cores through "personal feedback loops"; (2) center and elevate their leadership mindsets; (3) serve and build caring and competitive cultures; and (4) demonstrate and inspire ethical and disciplined behaviors.

In this paper, the author details an effective leadership cycle for achieving technical excellence on personal and organizational levels. More specifically, the author provides (1) data supporting the leadership cycle; (2) guidelines for identifying, developing, and sustaining leaders; and (3) leadership metrics.

The target audience for this paper is anyone interested in pursuing technical, process safety, business, and personal excellence.

1. Background and Purpose

I have worked for or with technical managers my entire 17-year professional career. I have also managed technical teams, business units, consulting companies. As I continue my professional journey with my second technical consulting start-up, I believe the primary role of a technical manager is to lead his/her organization to achieve and sustain technical excellence. The problem with this statement lies in the definition of "technical excellence." More specifically: (1) what is technical excellence?; (2) what are the specific components of its definition?; (3) how do you achieve technical excellence?; and (4) how do you measure technical excellence? The answers to these four questions are the deliverables of this paper.

During the Spring of 2012, my company, Cognascents Consulting Group, Inc., distributed a survey designed to gauge specific process safety leadership attributes. Upon analysis of the results, the following data and analysis surfaced:

1. 111 responses were received;
2. 40% of all respondents were technical managers or executives;
3. 32% of all respondents were process safety specialists;
4. 16% of all respondents were engineers;
5. The average age of all respondents was 47;
6. 59% of all respondents participate in a volunteer organization;
7. 93% of all respondents believe their friends and family would rate them as a safe driver;
8. 89% of all respondents were male;
9. 31% of all respondents had experience in the exploration and production industry, 25% in natural gas processing, 48% in refining, 62% in chemicals and petrochemicals, 41% in specialty chemicals;
10. 59% of all respondents identify themselves as conservative with respect to social philosophy, while 21% identify themselves as liberal;
11. Wisdom and Courage were considered the two most critical values for successful business leadership, while Temperance and Transcendence were considered the two least critical values for successful business leadership;
12. Wisdom and Courage were considered the two most critical values for successful business leadership, while Temperance and Transcendence were considered the two least critical values for successful process safety leadership;
13. 49% of all respondents felt business leaders are not held accountable for process safety performance the same way they are held accountable for business performance;
14. Random audits of a process safety management program are more effective than scheduled audits.

Specific survey data and results from the above list lend greater credibility to the content and proposals of this paper. More specifically, 40% of all respondents were technical managers or executives when the survey was administered. The average age of all respondents was 47 when the survey was administered, which represents the average age of the oil, gas, and chemicals industry given the ongoing Baby Boomer retirement effect. The pool of respondents had strong experience representation and diversity from across the oil, gas, and chemicals industries.

Although the survey was designed to derive greater insight into process safety leadership, the data and results coupled with additional research and self-study helped me better understand and define "technical excellence." After walking you through my derivation and definition of technical excellence, I will present an effective leadership cycle for achieving technical excellence on personal and organizational levels.

2. What Is Technical Excellence?

As defined by Merriam-Webster, "technical" is an adjective used to describe an entity as "having special and usually practical knowledge especially of a mechanical or scientific subject." Additionally, "excellence" is a noun meaning "a quality of being eminently good." Hence, Merriam-Webster would have us define "technical excellence" as the "quality of being eminently good at a special and usually practical knowledge, especially of a mechanical or scientific subject." When applied to the oil, gas, and chemical industries, this definition of "technical excellence" is fair and accurate, but incomplete. I have observed people who seemingly achieved the above definition of technical excellence. I too have seemingly achieved technical excellence according to Merriam-Webster. Before characterizing me as arrogant, let me further define "technical excellence."

Technical excellence is indeed a quality. As mentioned above, qualities are used to describe things. Some qualities, like hardness, can be intrinsic. Other qualities, such as tolerance, can be acquired or developed. Humans measure qualities against standards, benchmarks, or their opposites. Some qualities change and/or cycle over time. By applying this additional understanding of "excellence", I propose a second, modified definition of "technical excellence": an intrinsic and/or developed quality for a given state of existence of being eminently good at a special and usually practical knowledge as measured against a standard.

The above definition is more fair and more accurate, but still incomplete. Why? Because there is a virtuous component to "excellence." "Paideia" was the ancient Greek process of nurturing and developing young minds into the ideal standard of citizenship. Paideia was designed to achieve "arete", which translates into "virtue," but is intended to describe one's greatest human potential…or excellence. It is important to note that human potential to the Greeks is not the same thing as individual potential. Human potential implies effectiveness beyond oneself. This brings us to my final proposed definition of "technical excellence:"

> *An ever-transcending quality of being eminently good at a special and usually practical knowledge as measured against a standard of effectiveness defined by the possessing entity's community.*

So, what does this all mean? Technical excellence is a quality, a state of existence, a virtuous condition, a measurable outcome. However, technical excellence is not achieved once. It is achieved over and over again. The standard of effectiveness against which excellence is measured is dynamic; hence, the quality of technical excellence is dynamic. The state of technical excellence is dynamic. The virtuous condition of technical excellence is dynamic. The measurable outcome of technical excellence is dynamic. This dynamicism of technical excellence is what drives our continuous transcendence from a lesser to a higher state of technical performance. Achieving and sustaining technical excellence require constant evolution on our part (i.e. transcendence), which is difficult at times and often the reason for failure in pursuit of technical excellence. But why?

3. Organizational Leadership and Learning Model

One of the mantras of the ancient Greek paideia was "Know thyself." These two words are easy to spell, easy to say, and easy to understand. But they are hard to do. According to the ancient Greeks, knowing oneself was critical to maximizing one's human potential. Everything flows from your core. Your core centers and elevates you as a leader. As a leader, you serve and build your organization's culture. Your organization's culture influences the behaviors within your organization. Your organization's behaviors drive your outcomes. Outcomes teach and inform your core. This last step in the organizational leadership and learning model is where we transcend as individuals…and as organizations. Outcomes and results teach and inform our cores through the wisdom of experience. As we get older, experience becomes the critical element to transcending our core values, strengths, weaknesses, intrinsic motivation, natural and environmental limitations. Experience becomes the critical element to transcending our leaders, cultures, behaviors, and outcomes. Hence, experience is the critical element in transcending from a lesser to a higher state of anything…including technical excellence.

Figure 1: Organizational Leadership and Learning Model

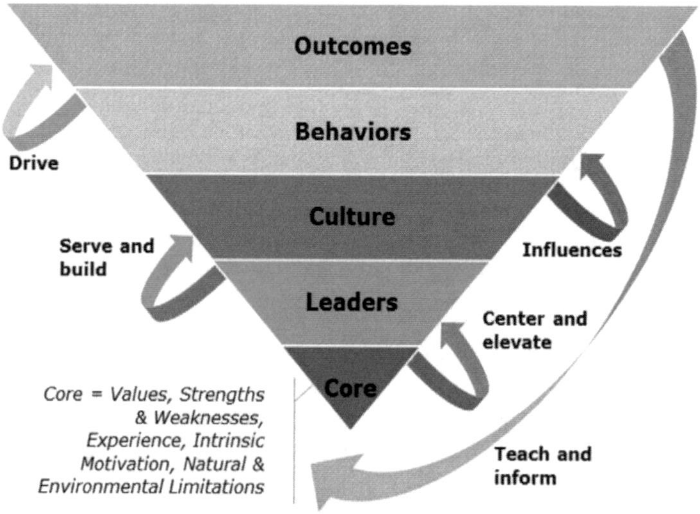

The above figure reveals how individuals within an organization impact the organization's performance. It tells us that transcendence is highly dependent on learning from our experiences. Put another way, failure to transcend implies a failure to learn from our experiences. Could this be true? Let us revisit specific data and analysis from the previously mentioned process safety leadership survey:

1. 93% of all respondents believe their friends and family would rate them as a safe driver;
2. Wisdom and Courage were considered the two most critical values for successful business leadership, while Temperance and Transcendence were considered the two least critical values for successful business leadership;
3. Wisdom and Courage were considered the two most critical values for successful business leadership, while Temperance and Transcendence were considered the two least critical values for successful process safety leadership;

One of the reasons for asking respondents to indicate how friends and family would rate their driving skills is to determine, albeit fairly subjectively, how honest respondents are with themselves. The fact that 93% of all respondents believe their friends and family would say they are good drivers may indicate a lack, to some degree, of self-awareness (i.e. we may not know ourselves as well as we think).

In addition, we excel at what we practice and we practice what is important to us. Transcendence and temperance were rated below all other choices as critical to business and process safety leadership success. The data suggests individuals within organizations do not understand the importance and experience-based mechanism of transcendence to achieving and sustaining technical excellence. To achieve and sustain technical excellence, leaders need to design and implement their own leadership feedback loops. That's right, a leader must design and implement a leadership feedback loop…it does not just happen.

4. The GLASS Revolution for Excellence

4.1 GLASS: A Revolution for Excellence

Since elementary school, "Excellent" has been a mark of distinction…an elusive grade for many, but always the standard to measure against. Before doing anything, you should ask yourself, "What do you want to achieve?" Given the subject and content of this paper, I am going to set the goal as *technical excellence.* Hence, your challenge is to achieve and sustain technical excellence. To sustain your pursuit of excellence, you should employ a cycle of actions. As I developed our cycle for excellence, I re-acquainted myself with one of the great, and often overlooked, technological discoveries: glass!

MacFarlane and Martin point out in their book, *The Glass Bathyscape*, how glass has become so commonplace in our lives that we often forget how valuable it is. Try for a minute to imagine a world without glass…no microscopes, telescopes, thermometers, contact lenses, stained-glass wonders, light bulbs, or high-speed driving! When you stop and think about it, glass is an amazing substance of balance and evolution. It is hard, yet brittle. It is transparent, but also reflects. It is created by cooling molten substances; hence it is a liquid and a solid during its lifetime. Glass continues to evolve even today through technological advances – making it thinner, more flexible, stronger. It seemed natural and fitting for our final cycle of excellence to mirror the balanced and evolutionary nature of glass. Our GLASS Revolution for Excellence requires transparent leadership and reflective learning, strong discipline and flexible strategies, hard metrics and sensitive adjustments.

Figure 2: GLASS Revolution for Excellence

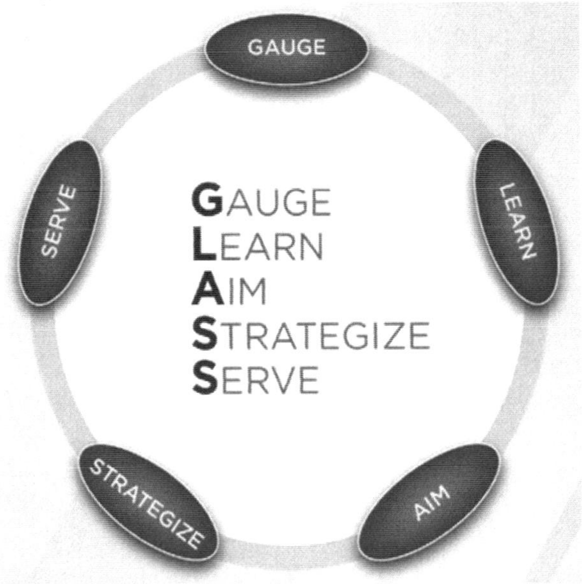

4.1.1 Gauge

The first action is to GAUGE. Before starting our journey to excellence, we must first figure out where we are and how we measure up against our personal expectations as well as external benchmarks and standards. This action is more than just measuring…it involves measuring against a standard of effectiveness. I propose benchmarking with caution as we do not want to chase the pack in pursuit of mediocrity. In order to gauge, we must first gather baseline information to help you identify what needs to be done for immediate and sustained excellence. The first step of gathering baseline information is where most people start deviating from the path to excellence. Gathering information is not what most people want to do when they are ready to take action. What do most people want to do when they are ready to take action? Anything…as long as they think it might help them achieve what they think they should be achieving. Only after gathering and analyzing information can we move to the second action of our revolution. According to the survey results, random audits are preferred to scheduled audits. I also recommend unscheduled on-demand competency checks for an objective measure of your organizations' competency level.

Figure 3: GLASS Revolution - Gauge

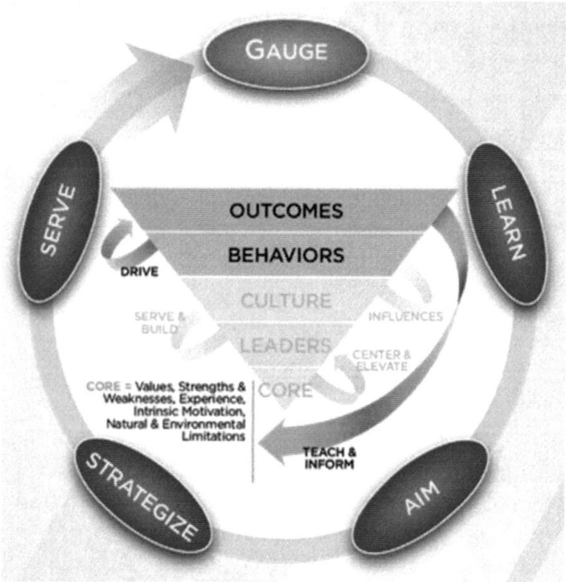

4.1.2 Learn

As previously mentioned, learning is the critical element to transcendence from a lesser to a higher state of excellence. Without establishing effective feedback loops for continuous "fine-tuning" of our cores, we as individuals will fail to evolve. I recommend developing personal feedback loops where outcomes and experiences are brought back to the core(s) of the leader(s) of your organization(s). Oftentimes, we design feedback loops where the outcomes and experiences are fed back directly to the Behavior layer of an organization. This is done by modifying a procedure or work process. This type of resolution does not drive individual, leadership, and cultural change. Outcomes and experiences need to be personally related to the leaders and expectations and clear lines of accountability need

Figure 4: GLASS Revolution - Learn

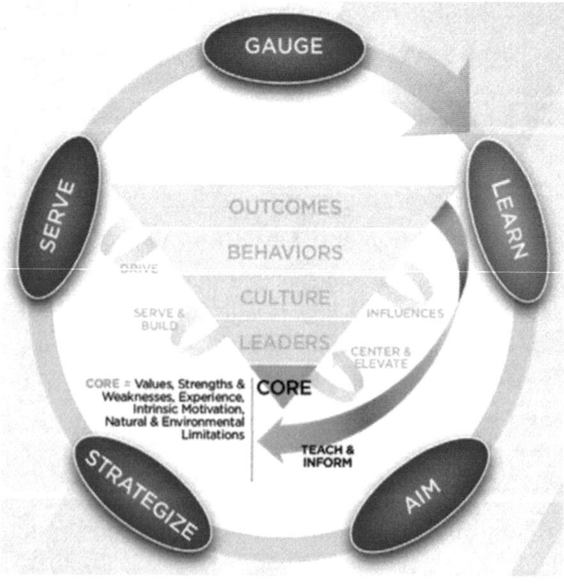

to be clarified. Leaders should be asked to personally re-state in his/her own words the lesson learned along with what he/she is going to do to practice and improve individual performance regarding the lesson learned. Identifying key performance metrics is best done at this stage of the cycle. Good questions to ask at this point in the process are (1) "how reliable is the measured outcome?" and (2) "is anything affecting the integrity of the measurement?" I have learned good leaders already have effective personal feedback loops in place. The most effective way to identify leadership talent is to look for individuals with effective personal feedback loops. These are individuals who are very self-aware and have already identifies their weaknesses and, more importantly, have already safeguarded their blind spots. Evaluation and survey of an organization's performance along with subjective feedback on the leadership in place can be effective. A behavior many organizations need to change is the practice of fast-tracking high-achievers with promotion after promotion. High-achievers or individual all-stars are often "pegged" as having high leadership potential; however, few are left in a leadership position long enough to learn and develop effective personal feedback loops to achieve and sustain technical excellence.

4.1.3 Aim

The third action is to aim. Aiming is different than strategy. Aiming is more about developing the vision of where to take yourself and the organization. With respect to technical excellence, aiming requires you to define what type of organization you want to be. Aiming also points out and identifies critical barriers to success. When you aim, you are not thinking at the 10 foot level of activity – you are thinking at the 10,000 foot altitude. Like a controller of a process control feedback loop, aiming calculates the difference between a measured parameter and a desired "setpoint" (this difference is called the "error value"). Once you know how far off you are from where you want to be along with which direction you need to go, you can move on to the next step of strategy.

Figure 5: GLASS Revolution - Aim

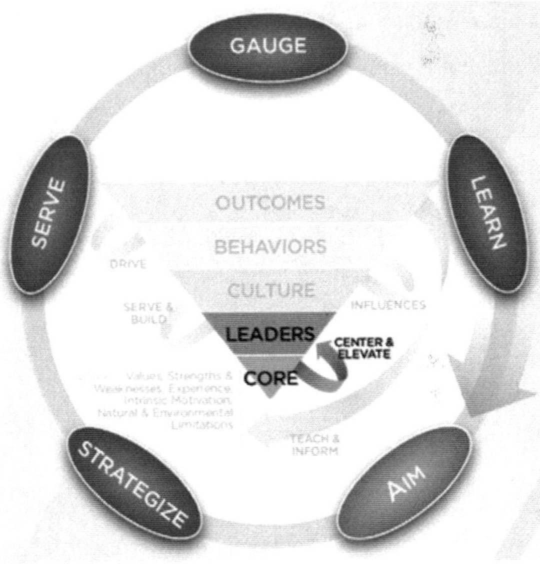

4.1.4 Strategize

The word "strategy" is of military origin and derives from the Greek words for army (stratos) and leader (agos). Strategy was first used to develop a plan of action to achieve victory on the field of battle. In the business world, the weapons are different and the battlefield is replaced by the marketplace, but strategy is still used by the "army's" leaders to plan for victory. Strategy is one responsibility never to be delegated by a leader. Leaders build and serve the cultures of their organizations by ensuring the strategy is cascaded throughout the organization using the following eight elements of an effective strategy:

1. Cascadeable – Each technical function must be addressed in the strategy. Each technical function should be able to take the strategy and apply it to their operations.
2. Adaptable – An ever-changing marketplace requires a dynamic and fluid technical and technology strategy.
3. Sustainable – A strategy must provide for sustainability through innovation, growth, and value relevance.
4. Competitive – A strategy must capitalize on existing competitive advantages or create new competitive advantages.
5. Achievable – A leader must balance the scope of the strategy with the depth of the organization's technical capability and capacity. A strategy must address retention and/or acquisition of resources and capabilities necessary for optimal execution.
6. Defensible – A leader must be able to defend all parts of the strategy with hard and current data. Risk assessments (identification and plans for elimination or mitigation) are must-haves for any strategy.
7. Evaluable – A leader must be able to monitor and measure the performance of the strategy against practical indicators.
8. Significant – A strategy must be relevant to the current company make-up (except for the rare conglomerates).

Figure 6: GLASS Revolution - Strategize

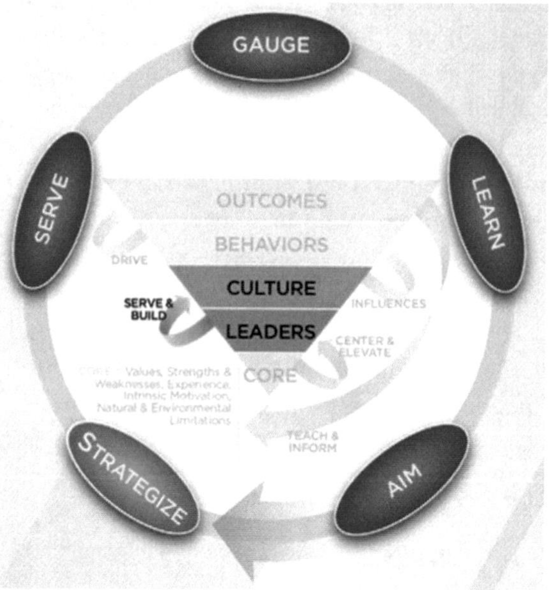

4.1.5 Serve

The fifth and last action is to serve. "Serve" describes how we should act once action is required. The organizational leadership and learning model is an inverted pyramid to demonstrate (1) the number of relationships and deliverables for each layer and (2) the hierarchy of service from bottom-up. Your core serves your leadership potential and performance. As a leader, you serve and build the culture of your organization. An organization's culture serves (influences) its communities and individuals by establishing behavioral norms, work processes, systems, and procedures. Lastly, organizational behaviors serve (drive) performance outcomes. A servant mindset and attitude allow us to maintain a sense of vulnerability as we humbly help others. A servant places others first and is always thinking

Figure 7: GLASS Revolution - Serve

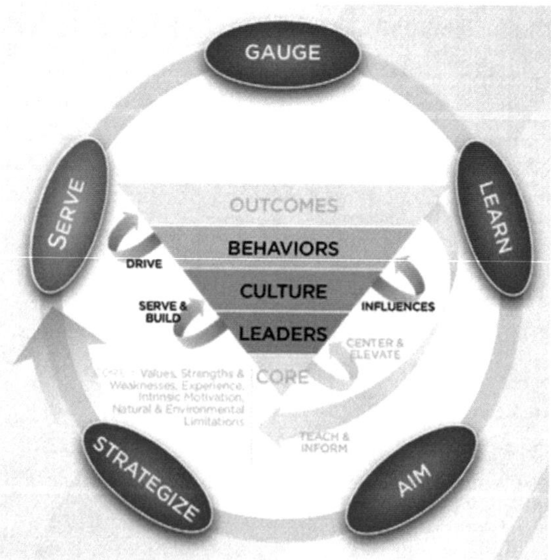

about their human potential in terms of impacting others, positively or negatively. Upon completion of one revolution on the GLASS cycle, we start a second revolution. At the onset of the second revolution, the Gauge stage is used to measure the outcomes of the first revolution and define the new standard of effectiveness to measure our technical excellence against.

7. Conclusion

The stated deliverables of this paper were to answer the following four questions:

1. What is technical excellence?
2. What are the specific components of its definition?
3. How do you achieve technical excellence?
4. How do you measure technical excellence?

Based on survey results, independent research, and self-study, I have proposed the following definition for "technical excellence:"

> *An ever-transcending quality of being eminently good at a special and usually practical knowledge as measured against a standard of effectiveness defined by the possessing entity's community.*

I further posit technical excellence is a quality, a state of existence, a virtuous condition, a measurable outcome. Additionally, the standard of effectiveness against which excellence is measured is dynamic; hence, the quality of technical excellence is dynamic. This dynamicism of technical excellence is what drives our continuous transcendence from a lesser to a higher state of technical performance. Achieving and sustaining technical excellence require constant evolution on our part (i.e. transcendence), which is difficult at times and often the reason for failure in pursuit of technical excellence. Transcendence is difficult because it is highly dependent on learning from our experiences, which is not easily accomplished without the use of personal feedback loops.

Figure 8: GLASS Revolution Integrated with Organizational Leadership and Learning Model

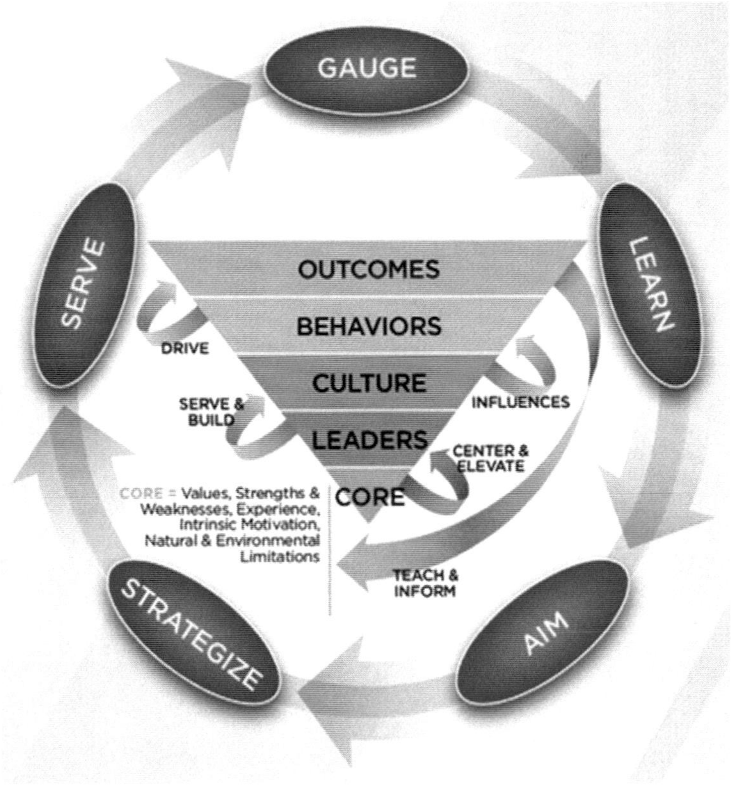

I fully acknowledge the need for further research to gather more information and more support for the positions and arguments I have made in this paper. Nonetheless, results from my initial survey suggest individuals within organizations do not understand the importance and experience-based mechanism of transcendence to achieving and sustaining technical excellence. Until more information and insight are obtained, I offer the GLASS Revolution for Excellence along with the Organizational Leadership and Learning Model to help individuals and organizations better identify, develop, and sustain leadership for technical excellence.

7. References

[1] Werner Jaeger, *Paideia: The Ideals of Greek Culture*, vols. I-III, trans. Gilbert Highet, Oxford University Press, 1945.

[2] Macfarlane, Alan and G. Martin. *The Glass Bathyscape*. Profile Books Ltd, 58A Hatton Garden, London ECIN 8LX. Copyright © Alan Macfarlane and Gerry Martin 2002.

(88b) Leading the Cultural Transformation: Achieving and Sustaining Excellence

Terry Mathis, ProAct Safety, The Woodlands, TX

Sir Winston Churchill observed, "First we shape our dwellings, and afterwards our dwellings shape us." Like dwellings, an organizational culture is either a powerful tool or a hindrance to shaping progress. Cultures influence the beliefs and behaviors of new members and should be managed for breakthrough performance in any operating category. Is your organization shaping the role culture plays in success, or are your dwellings determining this for you? Cultures can be understood, measured, and managed. The Safety Culture Excellence® Evolution Model has provided guidance for many complex organizations and has facilitated the transformation necessary for the acquisition of new performance. Many have written about the obvious role the beliefs and behaviors play in culture. The purpose of this talk, however, is to provide direction on how to lead, shape and sustain a culture of safety excellence.

Balanced Scorecard applicable to Textile projects in developing countries

Virgilio L. Gonzalez, PhD
AIChE : 119354
E-mail: gbasoc@cantv.net

SUMMARY

The textile industry is an area characterized by a productive chain of high employment generation. It includes Cotton crops, the manufacture of synthetic products with polyethylene and Polypropylene, needed to make yarns, the chemical treatment and manufacturing of fabrics, confection of garments and the trade of finished and intermediate products.

In spite of the competition of the Far East products, some countries in the Latin American areas have developed interesting profitable projects. As they cannot make some of their basic raw materials, including chemicals and machinery, there are opportunities for the USA to boost its trade.

The Balanced Scorecard (BSC) is a management tool developed more than ten years ago by economists Robert Kaplan and David Norton, that can be used as a guide for project managers, to help identifying opportunities for trade and consulting activities. In cases like textile manufacturing, the application of the BSC can help Developed and Developing countries to benefit.

In this work, a comparative Balanced Scorecard applied to the textile industry in two developing countries is summarized, showing perspectives and indicators that can help project managers optimize their work.

Introduction .- The Textile Industry

Textiles are present in practically every activity of life. Human beings dress and walk on textiles, sit down in chairs and sofas covered with fabrics, sleep on them and have even gone to the

Moon wearing more than twenty layers of special tailored fibers. Generally speaking, the Textile industry is responsible for the manufacture of basic dressing needs, fabrics, leather articles, footwear and many other goods commonly used.

The Textile industry is very complex. A typical textile chain consists of the preparation of fibers, being Cotton, silk, wool, linen, Nylon, Polyester and Polyethylene the most common to make yarn. Through Weaving, fabrics are made. Then, a Chemical treatment comes out to dye, finish and wash these fabrics. Finally, tailoring is carried out to get final products to the public. (See Figure No. 1).

FIGURE 1. TYPICAL TEXTILE CHAIN

```
FIBRES
Cotton, Linen, Silk,---→ YARN-→FABRIC-→DYING-→|
Wool,Polyester,                                |
Polyethylene, Nylon                            |
                                               /
TAILORING←- WASHING < --- FINISHING ←------   /
```

Each step can have many individual operations. For example, to make Yarn, one needs to take care of Cotton crops. Inevitably, as an agricultural product, quality can be affected by climatic conditions. Taken to plants, Cotton is prepared in packs. These packs go through an opening process to go to Unifloc machines. Flocs made go to a carding process, where straps result. Finally, through open end and ring machines among others, Yarn finally comes out. (Figure 2).

FIGURE 2. INDUSTRIAL MANUFACTURE OF YARN.

```
CROPS-→PACK ------→OPENING-→FLOC FORMATION→|
       PREPARATION                          |
                                            |
                                            /
                             YARN←---CARDING
```

The textile industry is also characterized for showing a very high employment index.

The features of the Textile Industry make it very complex to operate and manage, hence the need for special tools to have reasonable profitable results, particularly in present circumstances, due to the high competitiveness of Far East Countries.

The Scenario.

Latin American countries considered in this paper were Colombia and Venezuela.

For many years, Colombia has taken important steps to develop its Textile industry at different levels – Cotton crops, Industrial manufacture, Finishing and tailoring fabrics. It has also taken care of projecting their goods Worldwide and by joint efforts with the Governments, leading to important Free Trade Treaties with other Latin American countries, The USA, European Economic Community and Korea among others. Thanks to these factors, two of the largest Colombian mills, as are Fabricato and Manufacturas Eliot, are modernizing plants in more than 40%. Direct employment in textiles represents 12% of the Gross Domestic Product.

Venezuela has not got the same textile tradition as Colombia, but until 1990, it was its second most important source of employment. Due to Government interventions together with lack of orientated policies to developed textile industries, backed up by the fact that it is a major oil producer, importing goods from other countries at lower prices has resulted on the closing of major mills such as Sudamtex de Venezuela.

The Balanced Scorecard

The balanced Scorecard (BSC) is a managerial tool developed by economists Robert Kaplan and David Norton more than ten years ago. It allows company managers and directors to implement actions, in order to accomplish objectives through four perspectives: Financial, Customer orientation, Internal Processes, Innovation and Learning. TABLE 1 shows a Balanced Scorecard with special reference to a typical textile company in developing countries. Each perspective has its own strategic objectives and related indicators. Some results of the application of the Balanced Scorecard are now explained.

TABLE 1
BALANCED SCORECARD – TEXTILE INDUSTRY

Perspective	Strategic objectives	Indicators
Financial	Capital utilization	ROI,NPW,DCFR
	Cost reduction	% Cost raw materials
	Sales increase	% Machinery renovation
Customer	Satisfaction	%Satisfaction/production type
	Profitability	% Sales/customer
	Retention	% Faithful customers
Internal Process	Innovation	% Improvement operational logistics
	Operational process	% Plant stoppages
	Aftersales service	% Product devolutions
Innovation &	Employment capacity	Productivity
	Emp. Competitiveness	% Training
	Emp. Satisfaction	% Emp. Satisfaction
	Info Systems capacity	Competence level/Emp.

1.- **Financial perspective.**

1.1 Low returns on investment compared to other industrial sectors. Profits mainly come out on a production volume basis.

1.2 Seriously affected by cost of raw materials: Yarn and chemicals. Yarn manufacturing or purchasing very easily takes care of more than 60% of production cost. Colombia produces about 80% of Yarn needs.

1.3 Cost reduction strategies may be more relevant than sales increases.

1.4 Due to the need of foreign currency, seriously affected by imports. In the case of Venezuela, this is a most critical factor because of existing exchanged regulations.

2. **Customer orientation perspective**

2.1 More customer needs and requirements come out because of fashion and special needs for end users. This is most relevant in Colombia because this country is now differentiating as a key example for textiles manufacturing in textiles. They celebrate two of the most important fashion events in America, and because of free trade agreements, they import and export finished textile products Worldwide.

2.2 Customer more faithful to price than to product quality.

2.3 Tendency to import products, mainly from the Far East. Difficulties to compete with Textile goods from this part of the World.

2.4 Virtually all sales take place under long term credits.

3. **Internal process perspective**

3.1 Low tendency to innovate processes due to high operational costs and marketing uncertainty (fashion). However, some companies seek differentiation in finally tailored products to stay in the market- Some examples are:

- Manufacture of special fibers that allow transpiration without wetting fabrics.
- Making textile goods using waste products (rubber from wheels, PET, Doffing material after weaving, etc.
- Bullet proof fabrics

3.2 Tendency to give a good service. E.g. Service to many small customers instead of a few bigger ones.

3.3 Frequent bottlenecks found in industrial production and complex tailoring of finished products.

4.- Innovation and Learning Perspective

4.1 Low skilled working level. Lack of textile engineers. Venezuela has no relevant Textile Engineer Schools. Colombia has one but is not enough for the local needs of the Country.

4.2 High dependence of foreign skills to manage and maintain operating textile plants.

4.3 Tendency to buy used machinery to reduce investment / operating costs.

Opportunities for the USA Chemical Industry

A large amount of chemicals used in textile operations, as well as machinery and other raw materials can be exported to developing countries, such as the ones studies in this work. Most chemicals used to manufacture textiles are not made by developing countries.

In the case of Colombia, Free trade treaties could help USA companies making Sizing, Finishing and Printing chemicals to strongly get into this important Latin American Country, not only from a direct sales point of view, but also for the service and consultancy side. The Venezuelan case, in spite of the limitations, also sharing with Colombia the proximity of the North American market can also take the same advantages as Colombia. This is happening in some areas. For example, more than 70% of Yarn sizing chemicals used in Venezuelan textiles come from The USA.

The systematic application of Balanced Scorecards can contribute to better management and to make reasonable decisions about what to do next to keep Textile business alive in spite of current limitations.

Identifying the influence of low price textiles from the far East, together with the need for foreign imports, developing countries can try keeping business alive by differentiating actions. Some have been taken by countries like Colombia and Peru through Free Trade Treaties with the USA and Europe.

Developed countries manufacture textile machinery and chemical raw materials that can be exported to countries with free trade treaties with commercial advantages. Even countries where exchange regulations occur, have to purchase basic needs from developed countries to maintain industry alive.

In conclusion, the knowledge and development of Balanced Scorecards can definitely help project managers and directors to have a global vision of a company through its basic perspectives, and can motivate that projects in their different stages keep alive and operationally efficient.

Be an Inspired Safety Leader – Overcome Flaws that Cripple Effectiveness

Inspiration is not a word commonly used in safety programs, but it should be! Workers are asked to participate and cooperate and even get involved, but they are not asked to get inspired or to inspire others. We tend to concentrate on how to be safe rather than why it is important. We stress the numbers and overlook the people and lives that are damaged or lost. We see the money impact and forget the other critical ways that accidents impact operations and morale. We get hands and feet moving and neglect the hearts and minds.

It has been argued that the rationale for safety is obvious and that workers will participate simply out of self-interest. While this may be true of individual efforts, it falls far short when applied to the cultural aspects of safety. Excellence in safety performance involves the interactions of the group as well as the isolated actions of each individual. Workers look out for each other because they care, not because they are directed to do so. Remember Performance equals Skill times Motivation (P=SM). We tend to lead safety by increasing skill and overlooking the importance of motivation.

Many leaders who truly desire to improve safety make some common mistakes that undermine what they are trying to accomplish. Some actually damage their own safety programs while trying to improve them. Good intentions are not enough! You have to follow some simple guidelines and avoid the common flaws that can be detrimental to safety leadership efforts. A little attention to detail can produce drastically improved results. Learn the techniques that can make you a more inspired, inspiring and effective safety leader.

Learning Objectives:

- Discover the power of inspiring workers rather than simply directing them in safety
- Learn the most common flaws in safety leadership and discover how they damage, rather than improve, safety efforts
- Take away a list of do's and don'ts that will greatly improve your effectiveness
- Develop a vision of what safety excellence is and the business rationale for pursuing it
- Get inspired to become inspirational!

Using the Brains of Others to Accelerate Innovation

Many times when we are trying to innovate, we take the position that what we are trying to do is something that has never been done before. When we are trying to discover totally new science or engineering principles, this may be true, but in the normal realm of commercial innovation, this is seldom true. We often use fancy words, jargon, and acronyms to make us think our problems are special and unique. We even use these special words and phrases in web searches for solutions to the problems we have.

When we actually look at the inventive patent literature, we find a limited number of inventive principles that we constantly reuse and rediscover. This presentation will review a number of breakthrough innovation efforts in several manufacturing and engineering areas and their relationship to existing technology efforts that were overlooked and could have saved significant time and money in the innovation process.

STRATEGIC PLANNING MODEL, A CORE TOOL FOR INDUSTRIAL VISION.

Cruz Pedrero J. Manuel; Pintor Guzmán Pedro; Rodriguez Vega R. Alfonso; Reyes Guerra Bernardo; López de Cosío J. Alfredo; Lopez Molina M. Guadalupe.

Instituto Mexicano del Petróleo; Instituto Tecnológico y de Estudios Superiores de Monterrey; Universidad Popular Autónoma de Puebla; Universidad Iberoamericana Puebla.

mpedrero@imp.mx; ppintor@imp.mx; rrodriguez@imp.mx; breyesg@itesm.mx; josealfredo.lopez@upaep.mx; musi.lopez@iberopuebla.mx

ABSTRACT

Strategic planning, prospective strategic and innovation models have been used in all possible areas, such as military, civil, industrial, social, administrative, techniques, etc., so that the creative aspect of the vision has been core element to achieve particular objectives in the shortest time possible.

This paper describes some models of strategies used by growing businesses, as such as methodologies and planning tools that are the basis for the continued development of any industry or business.

Emphasis is made on models of known authors such as Michael Porter, Peter Senge, Robert S. Kaplan, his common tools such as 5 forces and Porter's diamond and Balanced Scorecard and others.

Finally we propose a model for strategic planning in marketing of chemical products used in the oil industry that improves the competitive vision of these products on the market.

INTRODUCTION

Many business models are based on ways traditional of implementation and strategic formulation that allow an increase in the non-disruptive nature of the business. Sven S. Volpel; Marius Leibold; Eden B. Tekie; Georg von Krogh (2003), so it is very difficult to stay in the competition and often do not survive.

The need to compete in a world in constant change and development has led that all the companies from the smallest to the largest changes and updates their activities both internal and external whose depend his performance on its economy, political and social stability.

Jaakko Heinola (2003) questioned what items are included in the creation of the strategy and portfolio management of the product and what kind of tools should be used to create product strategies in industry.

All of the above, coupled with the complex world of variables involved in the marketing of products and services there aren´t methodological model that ensures the timely delivery of products in quality and time, however relying on these variables and other models known, we can create organizational capacities well designed Allan Afuah (2003).

LITERATURE REVIEW

The following models of innovation and planning are an example of the experiences in the development of corporate models that have been designed and used to improve operation and performance all of each activities. I will mention any interesting models of authors[1-6] to support the importance of innovative companies.

Montealegre Model (2002)

Is the simplest model, was created to develop the capabilities of the company to support its business strategy. An electronics industry supporting on a vision of their resources created a strategy put a value on Stock Exchange of Venezuela, so it had to:

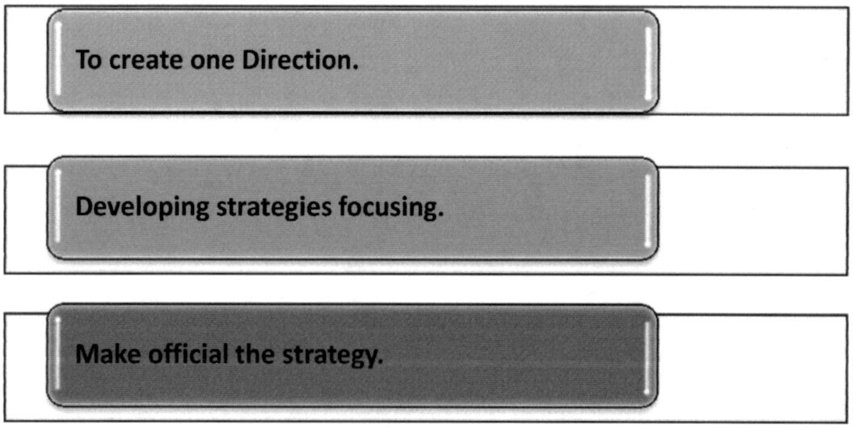

Resources in the process include: Leadership, organizational culture, technology of information, vision long-term and socials networks.

Hammel Model (2000)

Based on four principal components:

1. Capability people – abilties
2. Innovation measurement– metric.
3. Use of information technology – computing.

4. Redesign process – Administrative process.

Hammel suggests establishing a dynamic process: "Wheel of innovation.

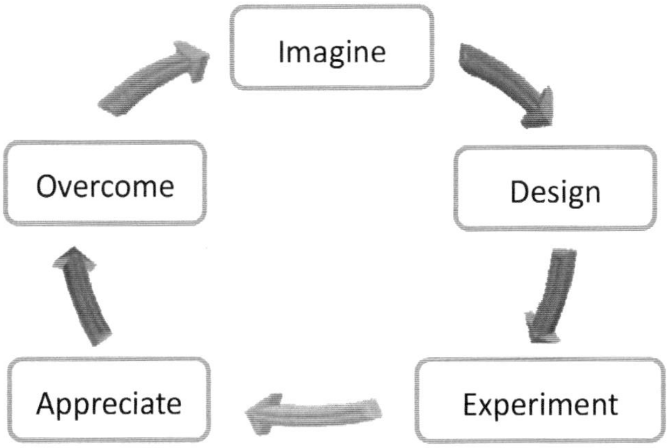

3M Model (1996)

In this model invert 7% of its budget on development and research to generating new products on the market with more than 30% of its sales. Is based on 6 principles:

1. Visión
2. Previsión
3. To extend the goals
4. Delegating
5. Comunications
6. Award and aknowledgement

People should be curious, eager to learn, to point to results, passionate, motivated by the achievement encourage first do it, then explain it and not be afraid to make mistakes.

HP Model

Have been deeply studied the characteristics of innovation in HP concluding in:

1. Qualified highly people.
2. Culture to accord to innovation.
3. Support to direction.
4. Interdisciplinary work in team

Carballo Model in Spain

Named "Model of benchmarking" is based on key words:

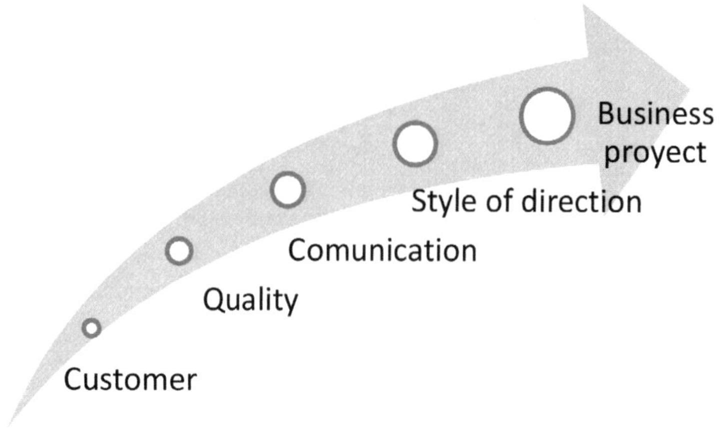

Is characterized because they are more agile in decision-making and avoid bureaucratic processes, its orientation is toward the market, the quality and efficient in information technology.

CEMEX Model

Based on its philosophy of constant innovation

1. Direction and leadership by Lorenzo Zambrano, cousin to founding member.
2. Knowledge depth of local market.
3. Results to point to customer.
4. Financial and operative control for products and services of world class to clients.
5. Passion for information technology results to point.
6. Ability to learn more quickly to the competence.

POLAR Model. Foods and beverages

Its main determinants after to make a depth study that included interviews, discussion groups and surveys were:

1. Orientation to market.
2. Management and leadership.
3. Knowledge of market.
4. Environment pleasant to innovation.
5. Apptitude to change.
6. Team working

Were analyzed a large number of variables to take advantage of the collected experiences obtaining the components:

1. Individuals.
2. Information.
3. Impulse.
4. Interaction.
5. Developing and research.
6. To point to market.
7. Intrepidity.

Led to the development of the potential capabilities to integrate, build and reconfigure internal and external competences to respond quickly to the changing environment to obtain strategic objectives in the review of its philosophy every five years.

Abernathy-Clark Model.

The following model considers the kind of innovation from of technological capacity one hand side and the ability to market with the other, to obtain the following:

Innovation	Technological capacity	Market capacity.
Regular	Maintain	Maintain
Niche	Maintain	Obsolete
Revolutionary	Obsolete	Maintain
Architectural	Obsolete	Obsolete

This model is static so it helps predict if innovation whether any company is preserved or tends to disappear.

Henderson Clark Model

This model involves 4 types of innovation based on the knowledge of components and knowledge of linkages between those or architectural.

Innovation	knowledge of components	knowledge architectural.
Incremental	Increase	Increase
Radical	Destroy	Destroy
Architectural	Increase	Destroy
Modulate	Destroy	Increase

This model complements the previous due to some companies have certain difficulties in its growth despite to obtain innovation type incremental.

Ro Model

5 avenues of decision

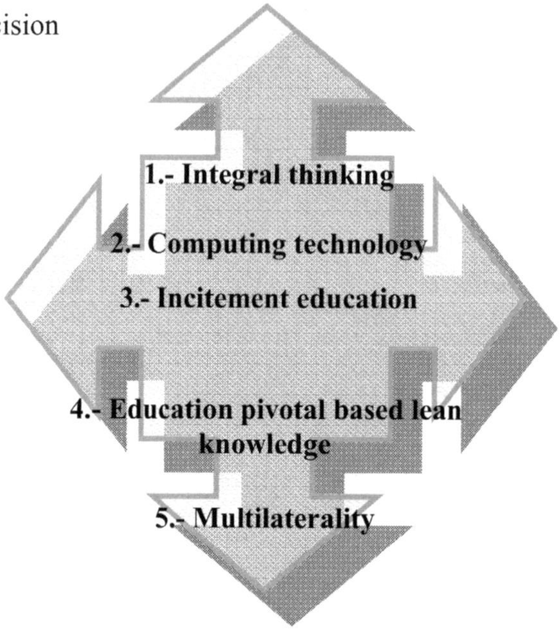

Peter Senge Model

This model set several concepts that relate service and human character within the company in all activities involved, production, sales, quality, etc.

1. Personal control.

   Arise, personal knowledge, autocontrol, love yourself
2. Thinking models.

   Avoid rare behavior and prejudice strongly influenced
3. Shared vision

   Avoid selfishness, apathy, jealousy and allow thinking free.
4. Team learning.

   Anyone is perfect and expert at all.
5. Sistemic focus

   Future vision with thinking developed in group.

Google´s model

Google's tetra helix is an example of the relationship between infrastructure (Internal perspective), customers (customer perspective), supply (perspective to improve and create value), finance (financial perspective) from Balanced Scorecard, described by Robert S. Kaplan[7] for companies with performance measurements.

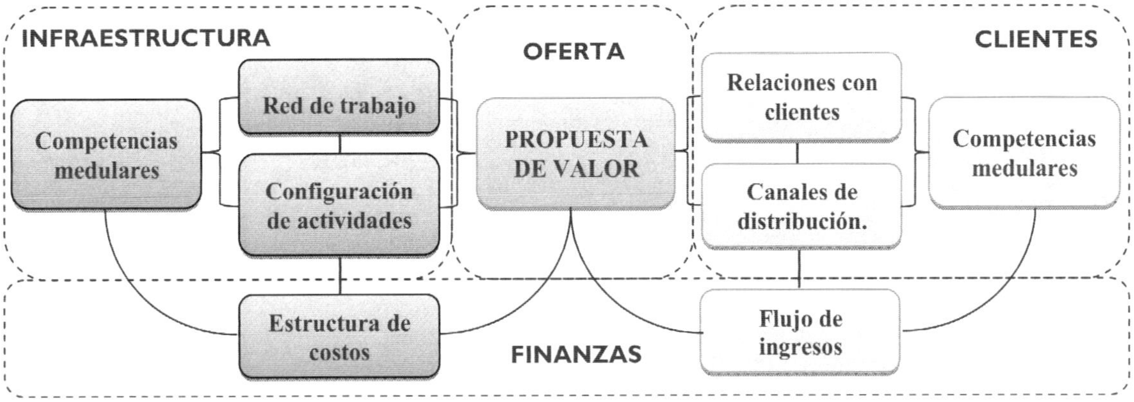

[Osterwalder (2004) The Business Model Ontology]

Chemicals Treatments Model

The current model seems complex at first glance but allows a relationship with suppliers, customers, and related support businesses, competition and threat of new products, partnerships with other agencies, solid infrastructure (patents and services), etc.

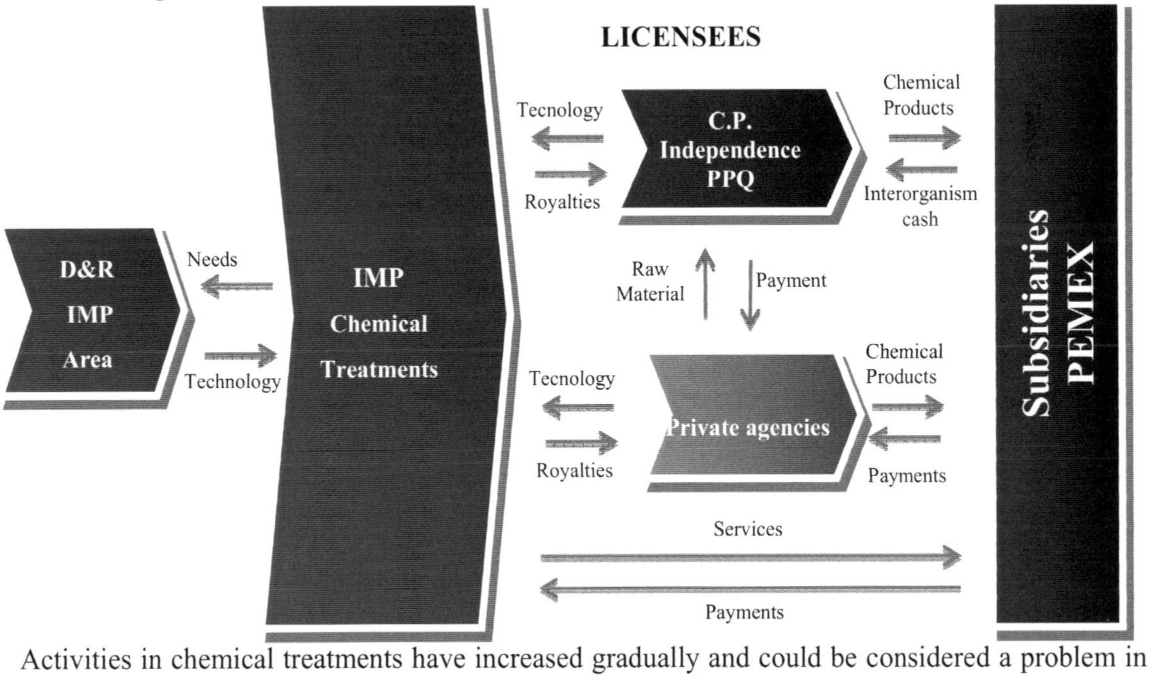

Activities in chemical treatments have increased gradually and could be considered a problem in the future if not taken into account any necessary changes to this model.

PROPOSE MODEL

We suggest a model to simplify the current activities and eliminate unnecessary steps than justify the savings for delivery timely of chemicals products in different consumption centers, and create value in specific functions of strategic character.

Today models work better than in the past but we need to make a planning creative activities for the functional performance at the same, is to say, for development of a technology, product or service your procedure in a timely manner should avoid duplication of activities, excess inventory overruns, higher costs of products in the market, etc. they may affect the stability of the company eventually.

Propose model builds on the structure of Michael Porter five forces, diamond pattern and integrates the value proposition and performance measures considering the following functions:

1. - Creation of the office of planning and strategic prospective in chemical treatments.

The following activities will help to chemical treatment area:

- Technological prospective analysis
- Technological networks global research for strategic products
- Costing raw materials and importation products
- Application IMP technology worldwide
- Vision and strategic objectives at least every five years.
- Updating and implementation of confidentiality agreements, contracts, copyrights, trademarks, patents, etc.

2. - Optimum methodology for marketing of technologies

- National and international market analysis
- Developing technological packages
- Competitive price of finished product
- Process flow diagram per customer. PEMEX, international, own.
- Method analysis even internal policy change

3. - Continue Training to produce creativity

4.- Organization, Leadership and Human Being Integration

IMP Model.

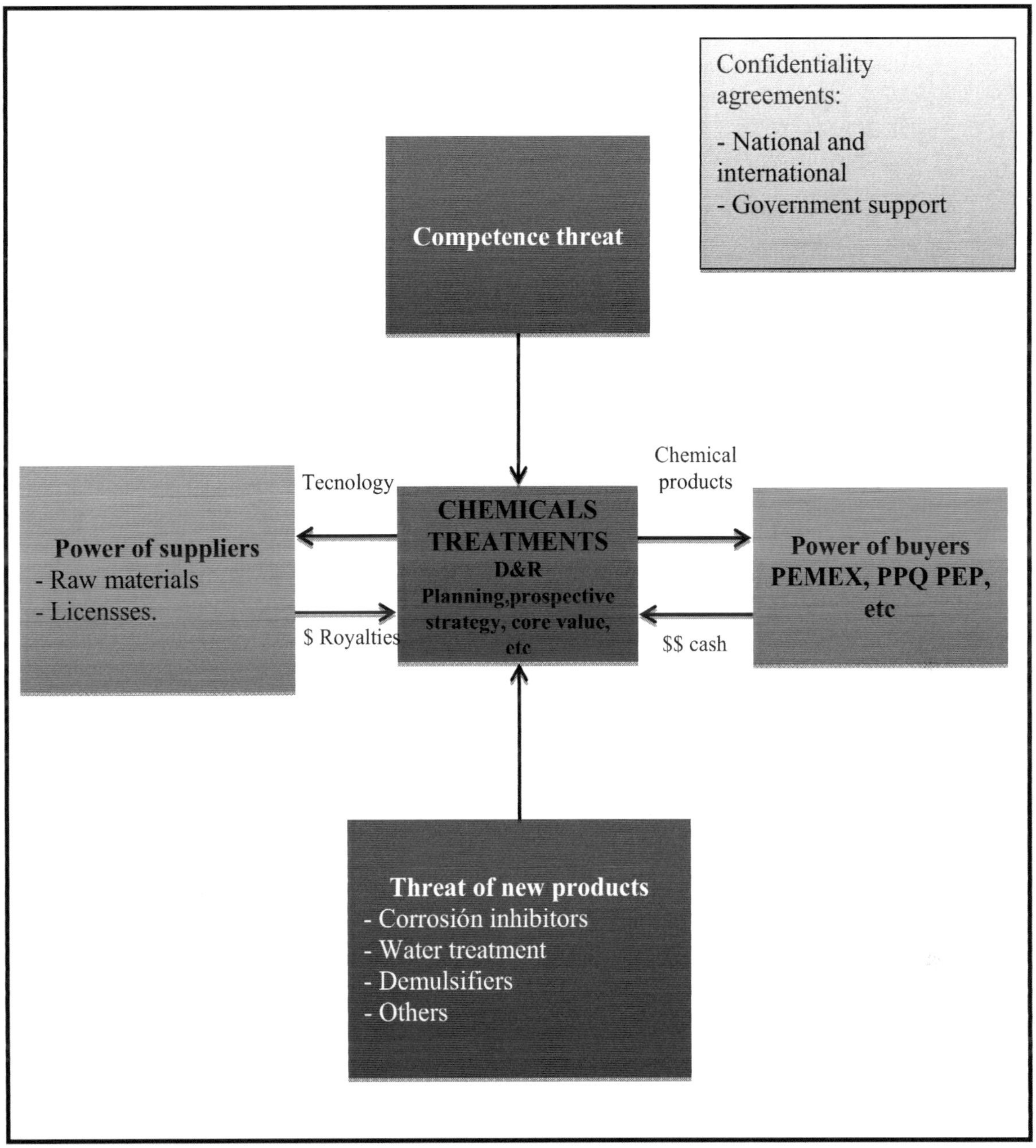

CHEMICALS TREATMENTS VISION

To be a strategic area in development, implementation, technologies and services integration in chemical treatments for national and international oil industry to ensure delivery of reliable and high quality products.

CONCLUSION

The model along with the methodology and the vision to integrating technologies it will lead align and arrange short-term activities, ensuring the competitiveness of the area of chemical treatments in the oil industry.

BIBLIOGRAPHY

1. Alberto Silva Aristeguieta. Un modelo para la innovación en la empresa. Anales de la Universidad Metropolitana. Vol. 5 No. 2 (75-78), 2005
2. Allan Afuah, "La dinámica de la Innovación Organizacional, el nuevo concepto para lograr ventajas competitivas y rentabilidad" Oxford University Press, 1999.
3. Michael Porter. The five competitive forces that shape strategy. Harvard Business Review. January 2008.
4. Michael Porter. The competitive advantage of nations. Harvard Business Review. March-April 1990.
5. Bernardo Reyes Guerra. Modelo ro. Apuntes curso de estrategias tecnológicas UPAEP primavera 2012.
6. Javier M. Cabanas G. Apuntes sobre el libro: "La Quinta Disciplina", Peter Senge. UPAEP primavera 2012.
7. El Balanced Scorecard: Mediciones que impulsan al desempeño. Robert S. Kaplan y David P. Norton. Lo mejor de Harvard Businnes Review. Julio 2005
8. Allan Afuah (2003) *Innovation management. Strategies, implementation and profits.* Oxford University Press. (2 ed) 13.
9. Jaakko Heinola. (2011). Developing a Systematic Model for Product Strategy and Product Portfolio Management. Master of Science Thesis Degree programme in industrial engineering and management. Tampere University of Technology.
10. Sven S. Volpel; Marius Leibold; Eden B. Tekie; Georg Von KrogH (2003). Escaping the red Queen effect in competitive strategy- how managers can change industry rules by Sense-Testing their business models.Institute of Innovation Research. Hitotsubashi University. Tokio Japon.

(115c) Expert Systems for Process Unit Troubleshooting

Blaise Arena, UOP, Des Plaines, IL and Alan Zagoria, CEOC, Des Plaines, IL

The need to document and disperse expert knowledge and experience throughout a large organization is an important challenge in most industries. At UOP we were challenged with an expanding and increasingly globalized Technical Services function. This group of people provides support services for thousands of our licensed process units in the worldwide refining and petrochemical industry. They diagnose and solve customer operating problems while relying on a comparatively smaller group of process specialists with deep knowledge of selected process technologies. Our objective was to broaden 24/7 access to the expertise of these specialists more widely across our organization. In addition, it was important to increase the troubleshooting and diagnostic capability of our less-experienced field personnel and improve our ability to respond to customer needs. We decided to develop a series of computer diagnostic tools using an artificial intelligence technique commonly called Expert Systems. These programs are designed to address troubleshooting needs for some of our licensed technologies. Over seven years, we have completed seventeen expert systems and implemented them for internal and customer use. These systems address process areas such as hydrocarbon separations, FCC, reforming, hydrogen purification, HF alkylation, propane dehydrogenation and catalyst selection.

Expert systems are interactive programs designed to prompt the user through a systematic decision making process, where the same decision choices must be made repeatedly. An expert system can be effectively applied to situations requiring diagnosis and problem solving. The interactive nature of the system simulates a conversation between a technology expert and the user. The system assists the user in troubleshooting the process unit operating problems and provides the necessary background information to make informed decisions. During a session with the expert system the user benefits by having the problem narrowed down, relevant information has been gathered and the problem may be solved outright. In addition, the user has learned from the interactive experience. The expert system is also a valuable teaching and learning tool.

Our approach to development of an expert system requires one or more technical experts to work in concert with a development and knowledge acquisition team. Through the knowledge acquisition process consideration is given to all relevant operational situations, generating discussion of special cases that might otherwise be overlooked. The expert systems approach allows the full complexity of an expert's decision-making process to be described and documented. It is especially helpful in complex situations involving balancing many interrelated factors or judgment-based decisions. The rule-based expert system approach, using heuristic rules processed by the expert system software inference engine, allows even very complex logic to be captured completely.

Providing 24/7 access to this level of troubleshooting expertise to our staff has provided benefits as mentioned above. We now also offer these expert systems to our customers on a subscription basis. Refinery process engineers or managers can directly access the expert systems online and begin immediate troubleshooting of a process unit upset. This initial effort can facilitate further discussion with our Technical Services staff, making the UOP-customer consultation more efficient.

The expert systems approach is an effective way to document organizational expertise and extend it to a broad user group who can exploit it to increase their own knowledge and capability. This approach can be used in any area that would benefit from expert guidance in decision-making. In this paper we will review the background, development principles, and benefits of the UOP Expert Systems project.

IndustryWeek
IW
IndustryWeek.com
LEADERSHIP IN MANUFACTURING

Failing Less is Not a Safety Strategy

If your safety program is based on lowering injury rates, it's time to think differently.

Wednesday, January 18, 2012
By

Since the inception of OSHA in the early 1970s we have measured safety by our failure rate. Like a golf score, we want it to be as low as possible and want it to be getting lower with each passing year. However noble a goal this might be, it is *not* a safety strategy. Yet, when I ask executives what their strategy is, the second most common response is to quote a targeted rate or reduction of rate (second only to mentioning a new program to be implemented).

Lowering a recordable rate is not a strategy, and a new program should only be part of a strategy. If you were launching a new product line or going for an increase in market share, you would develop a strategy to address the key issues impacting the desired results. Your strategy would address leadership, key skills needed, involving the right people, motivating those people, and measuring the right things to monitor success. So, rather than targeting the safety failure rate, how about targeting some issues that could impact the rate?

Below, I identify some of the targets that might help toward the development of a true improvement-based safety strategy.

1. Safety Leadership Skills – When the leaders lead, the followers follow. Do your organizational leaders lead safety like they lead other business activities and priorities? Do they establish the priority or value of safety to the organization and reinforce that through constant communication? Do they know how to lead safety or do they need help? Are the key leaders aligned on their approach to safety, and does the workforce get a unified message regardless of the messenger delivering it? Addressing and improving any of these aspects of leadership (and many more) could be a part of a good safety strategy and could make a significant impact on the organization's safety performance.

2. Performance Coaching – Has your organization discovered yet that safety is something your employees do, and that you can't do it for them or stand over them constantly and make them do it? Your managers and supervisors are coaches, not players. Their primary tool for improvement is to help others improve. We call this "coaching." Unfortunately, even though coaching is one of the most important skills for leaders, few leaders have had any formal training for coaching. A little formal training can go a long way toward safety improvement. But training is only a beginning and not a complete strategy. The training needs follow-up, and the use of the new skills needs to be reinforced and evaluated over time to ensure that it becomes a permanent part of the culture. It needs to find its way into job descriptions or RREs (Roles, Responsibilities and Expectations) for key leaders and supervisors. It needs to become a part of employee evaluations and new-employee orientation.

3. Employee Involvement Opportunities – The more active employees are in safety efforts, the better they work. This, of course, depends on the quality of involvement not just the time spent. Many organizations want more involvement, but fail to provide the structure for it to happen. What can employees join, do, volunteer for, or get involved in? There is no magic answer. For some organizations it is OSHA's Voluntary Protection Program, for some it is formal safety programs, and for others it is audit teams or safety blitzes. Whatever structure you choose, make sure it can succeed. Ensure that results of the activities are highly visible, and that participation is rewarded and not inadvertently or informally punished. Even a frown from the boss when you return from participating in a safety activity can doom the participation.

4. Improved Safety Communications – Is your safety communication delivered via multi-media? Do you measure the reception of the information or just the sending? Does your safety communication create focus and improvement or just cause information overload? My colleague Shawn Galloway developed a metric for several organizations he calls ROA (return on attention). When you ask for people's attention and send them a message, what do you get in return? How "sticky" are your messages? Are they memorable or do they get forgotten quickly? Improving safety communication can be a very worthwhile part of an overall safety strategy. You don't have to do everything at once. Many organizations address one or two aspects of safety communication per year and target specific improvements.

5. Revised Safety Motivational Program – Many safety incentive plans are not just ineffective; they are dangerous. With the wealth of new information available on what truly motivates workers, it is definitely time to review your strategy for safety motivation, or consider implementing one if you don't already have one. Review the new literature (including the free white papers available online) and take advantage of the new approaches available. Warning: Be careful not to make your change de-motivating by taking away rewards or incentives. Suddenly stopping the old program, even to start a new one, can create the illusion of a take-away. Rather, transition your existing program to a more effective one using the same budget.

6. Improved Safety Metrics – If you want to improve the lagging indicators (and you certainly do), discover the leading indicators that predict the improvements. If you choose to pursue one of these six suggestions, measure your progress toward the goal and see if the goal actually impacts the safety performance. This goal can help with goals #4 and #5 as well. Communicate the new metrics as a part of your overall safety communication improvement strategy and you will find that "visible improvement" is one of the new tools for motivation. Setting and achieving goals can help build a "can do" safety culture. What better quality for a culture than the ability to target and accomplish improvement?

All safety strategies begin with thinking differently. All safety improvement comes from doing things differently. The most effective strategies target the most transformational issues, i.e. the issues that can make the biggest differences in safety performance. Look at these suggested targets and honestly rate your organization's performance in these areas. Where are the weaknesses and where are the opportunities to improve performance the most? This is strategic thinking. Use it to create an improvement-based strategy that will enhance your safety performance. When you succeed more, by default, you fail less!

Terry L. Mathis is the founder and CEO of ProAct Safety, and was recently named one of 'The 50 People Who Most Influenced EHS' for the 2nd consecutive year by EHS Today. He can be reached at 800-395-1347 or info@proactsafety.com.

See also:

AUTHOR INDEX